Encyclopedia of Optical Fiber Technology: Selected Concepts

Volume IV

Encyclopedia of Optical Fiber Technology: Selected Concepts Volume IV

Edited by **Marko Silver**

New York

Published by NY Research Press,
23 West, 55th Street, Suite 816,
New York, NY 10019, USA
www.nyresearchpress.com

Encyclopedia of Optical Fiber Technology: Selected Concepts
Volume IV
Edited by Marko Silver

International Standard Book Number: 978-1-63238-148-4 (Hardback)

Printed in the United States of America.

Contents

Preface

This book focuses on current research and new advances in the field of optical fiber technology. It covers variety of topics about different aspects of the technology. The book includes study in the fields of new optical fibers and sensors. It aims to facilitate industrial users and researchers in their research activities by highlighting ongoing developments in optical fiber applications, fiber lasers and sensors.

The information shared in this book is based on empirical researches made by veterans in this field of study. The elaborative information provided in this book will help the readers further their scope of knowledge leading to advancements in this field.

Finally, I would like to thank my fellow researchers who gave constructive feedback and my family members who supported me at every step of my research.

Editor

Part 1

New Optical Fibers

Fabrication and Applications of Microfiber

K. S. Lim[1], S. W. Harun[1,2], H. Arof[2] and H. Ahmad[1]
[1]Photonic Research Center, University of Malaya, Kuala Lumpur,
[2]Department of Electrical Engineering, Faculty of Engineering, University of Malaya,
Kuala Lumpur,
Malaysia

1. Introduction

Microfibers have attracted growing interest recently especially in their fabrication methods and applications. This is due to a number of interesting optical properties of these devices, which can be used to develop low-cost, miniaturized and all-fiber based optical devices for various applications (Bilodeau et al., 1988; Birks and Li, 1992). For instance, many research efforts have focused on the development of microfiber based optical resonators that can serve as optical filters, which have many potential applications in optical communication and sensors. Of late, many microfiber structures have been reported such as microfiber loop resonator (MLR), microfiber coil resonator (MCR), microfiber knot resonator (MKR), reef knot microfiber resonator as an add/drop filter and etc. These devices are very sensitive to a change in the surrounding refractive index due to the large evanescent field that propagates outside the microfiber and thus they can find many applications in various optical sensors. The nonlinear properties of the micro/nanostructure inside the fiber can also be applied in fiber laser applications. This chapter thoroughly describes on the fabrication of microfibers and its structures such as MLR, MCR and MKR. A variety of applications of these structures will also be presented in this chapter.

2. Fabrication of microfiber

2.1 Flame brushing technique

Flame brushing technique (Bilodeau et al., 1988) is commonly used for the fabrication of fiber couplers and tapered fibers. It is also chosen in this research due to its high flexibility in controlling the flame movement, fiber stretching length and speed. The dimension of the tapered fiber or microfiber can be fabricated with good accuracy and reproducibility. Most importantly, this technique enables fabrication of biconical tapered fibers which both ends of the tapered fiber are connected to single-mode fiber (SMF). These biconical tapered fibers can be used to fabricate low-loss microfiber based devices.

Fig.1 shows a schematic illustration of tapered fiber fabrication based on flame brushing technique. As shown in Fig. 1, coating length of several cm is removed from the SMF prior to the fabrication of tapered fiber. Then the SMF is placed horizontally on the translation stage and held by two fiber holders. During the tapering, the torch moves and heats along the uncoated segment of fiber while it is being stretched. The moving torch provides a uniform heat to the fiber and the tapered fiber is produced with good uniformity along the heat region. To monitor the transmission spectrum of the microfiber during the fabrication, amplified spontaneous emission (ASE) source from an Erbium-doped fiber amplifier (EDFA) is injected into one end of the SMF while the other end is connected to the optical spectrum analyzer (OSA). Fig. 2(a) shows diameter variation of the biconical tapered fiber fabricated using the fiber tapering rig while Fig. 2(b) shows the optical microscope image of the tapered fiber with a waist diameter of 1.7μm. With proper tapering parameters, the taper waist diameter can be narrowed down to ~800nm as shown in Fig. 2(c).

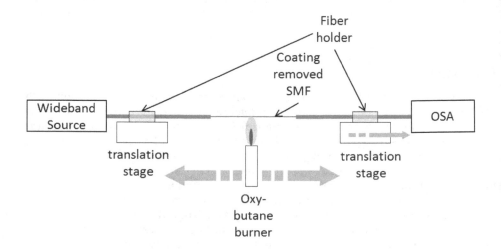

Fig. 1. Tapered fiber fabrication using flame brushing technique.

Adiabaticity is one of the important criteria in fabricating good quality tapered fibers. It is commonly known that some tapered fibers suffer loss of power when the fundamental mode couples to the higher order modes. Some fraction of power from higher order modes that survives propagating through the tapered fiber may recombine and interfere with fundamental mode. This phenomenon can be seen as interference between fundamental mode HE_{11} and its closest higher order mode HE_{12}. This results to a transmission spectrum with irregular fringes as shown by the dotted graph in Fig. 3 and the excess loss of the tapered fiber is ~0.6dB (Ding et al., 2010; Orucevic et al., 2007). This tapered fiber is not suitable to be used in the ensuing fabrication of microfiber devices. The solid curve in the same figure shows the transmission of a low loss tapered fiber with approximately more than 4mm transition length and the insertion loss lower than 0.3dB. Some analysis suggests that the coupling from fundamental mode to higher order modes can be minimized by optimizing shape of the tapers. In practice, adiabaticity can be easily achieved by using sufficiently slow diameter reduction rate when drawing tapered fibers or in other words manufacture tapered fibers with sufficiently long taper transition length. A detail discussion on the adiabatic criteria and optimal shapes for tapered fiber will be presented in the next section.

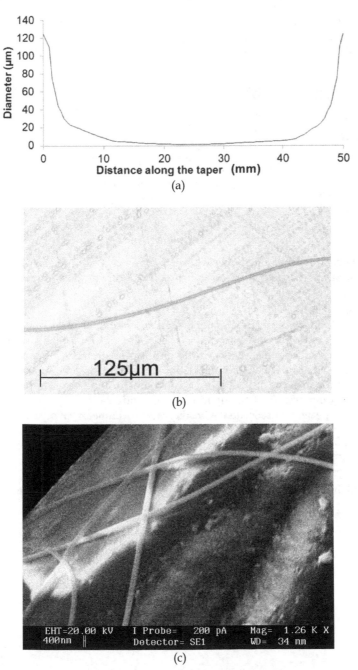

Fig. 2. (a)The diameter variation of a biconical tapered fiber fabricated in the laboratory (b) Optical microscope image of tapered fiber with a waist diameter of 1.7 μm (c) SEM image of a ~700nm waist diameter tapered fiber.

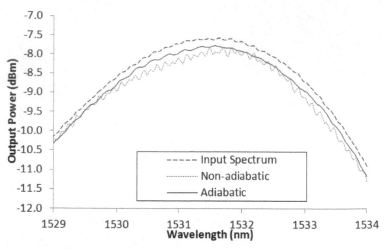

Fig. 3. Output spectra from a microfiber with 10 cm long and ~3μm waist diameter. Input spectrum from EDFA (dashed), adiabatic taper (solid) and non-adiabatic taper (dotted).

2.2 Adiabaticity criteria

Tapered fiber is fabricated by stretching a heated conventional single-mode fiber (SMF) to form a structure of reducing core diameter. As shown in Fig. 4, the smallest diameter part of the tapered fiber is called waist. Between the uniform unstretched SMF and waist are the transition regions whose diameters of the cladding and core are decreasing from rated size of SMF down to the order micrometer or even nanometer. As the wave propagate through the transition regions, the field distribution varies with the change of core and cladding diameters along the way. Associated with the rate of diameter change of any local cross section, the propagating wave may experience certain level of energy transfer from the fundamental mode to a closest few higher order modes which are most likely to be lost. The accumulation of this energy transfer along the tapered fiber may result to a substantial loss of throughput. This excess loss can be minimized if the shape of the fabricated tapered fiber follows the adiabaticity criteria everywhere along the tapered fiber (Birks and Li, 1992; Love et al., 1991).

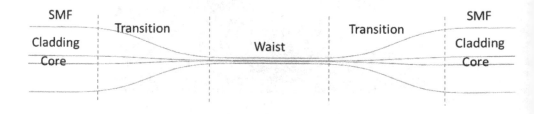

Fig. 4. Typical diameter profile of a tapered fiber.

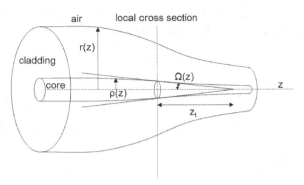

Fig. 5. Illustration of the taper transition.

Fig. 5 gives an illustration of a tapered fiber with decreasing radius where z denotes the position along the tapered fiber. Theoretically, an adiabatic tapered fiber is based on the condition that the beat length between fundamental mode LP_{01} and second local mode is smaller than the local taper length-scale z_t.

$$z_b > z_t \tag{1}$$

Referring to illustration in Fig. 5, z_t is given by

$$z_t = \rho / \tan \Omega \tag{2}$$

where $\rho = \rho(z)$ is the local core radius and $\Omega = \Omega(z)$ is the local taper angle. The beat length between two modes is expressed as

$$z_b = \frac{2\pi}{\beta_1 - \beta_2} \tag{3}$$

where $\beta_1 = \beta_1(r)$ and $\beta_2 = \beta_2(r)$ are the propagation constants of fundamental mode and second local mode respectively. From the above equations, Inequality (1) can be derived to

$$\left| \frac{d\rho}{dz} \right| = \tan \Omega < \frac{\rho(\beta_1 - \beta_2)}{2\pi} \tag{4}$$

where $\frac{d\rho}{dz}$ is the rate of change of local core radius and its magnitude is equivalent to $\tan \Omega$.

For the convenience of usage and analysis, Inequality (4) is rewritten as a function of local cladding radius $r = r(z)$,

$$\left| \frac{dr}{dz} \right| < \frac{r(\beta_1 - \beta_2)}{2\pi} \tag{5}$$

Based on this condition, adiabatic tapered fiber can be acquired by tapering a fiber at a smaller reduction rate in diameter but this will result to a longer transition length. Considering practical limitations in the fabrication of fiber couplers or microfiber based devices, long tapered fiber may aggravate the difficulty in fabrication. For the purpose of miniaturization, short tapered fiber is preferable. To achieve balance between taper length and diameter reduction rate, a factor f is introduced to Inequality (5) and yields

$$\frac{dr}{dz} = -\frac{fr(\beta_1 - \beta_2)}{2\pi} \tag{6}$$

where the value of f can be chosen between 0 to 1. Optimal profile is achieved when $f = 1$. Practically, tapered fiber with negligibly loss can be achieved with $f = 0.5$ but the transition length of the tapered fiber is 2 times longer than that of the optimal tapered fiber.

2.3 Shape of tapered fiber

When a glass element is heated, there is a small increment in the volume under the effect of thermal expansion. However, the change in volume is negligibly small not to mention that the volume expansion wears off immediately after the heat is dissipated from the mass. It is reasonable to assume that the total volume of the heated fiber is conserved throughout the entire tapering process. Based on this explanation, when a heated glass fiber is stretched, the waist diameter of the fiber is reduced. The calculation of varying waist diameter and length of extension can be made based on the idea of 'conservation of volume'. Birks and Li (1992) presented simple mathematical equations to describe the relationship between shapes of tapered fiber, elongation distance and hot-zone length. Any specific shape of tapered fiber can be controlled by manipulating these parameters in the tapering process. The differential equation that describes the shape of the taper is given by

$$\frac{dr}{dx} = -\frac{r}{2L} \tag{7}$$

where L denotes the hot-zone length and r denotes the waist diameter.
The function of radius profile is given by the integral

$$r(x) = r_0 \exp\left(-\frac{1}{2}\int\frac{dx}{L}\right) \tag{8}$$

To relate the varying hot-zone length L with the elongation distance x during the tapering process, L can be replaced with any function of x. Linear function

$$L(x) = L_o + \alpha x \tag{9}$$

makes a convenient function for the integral in Eqn (8).

$$r(x) = r_0\left(1 + \frac{\alpha x}{L_0}\right)^{-1/2\alpha} \tag{10}$$

where r_0 denotes the initial radius of the fiber. To express the taper profile as a function of z, distance along the tapered fiber is given as;

$$r(z) = r_0\left(1 + \frac{2\alpha z}{(1-\alpha)L_0}\right)^{-1/2\alpha} \tag{11}$$

By manipulating the value of α, several shapes of tapered fiber can be produced such as reciprocal curve, decaying-exponential, linear and concave curve. Several examples of calculated taper shape based on different values of α can be found in the literature of (Birks and Li, 1992). Consider the case of tapered fiber with decaying-exponential profile as shown in Fig. 6, the fabrication of such tapered fiber requires a constant hot-zone length (α=0). From the theoretical model presented above, the function for the decaying-exponential profile is given by

$$r(z) = r_0 \exp(-z / L_0) \tag{12}$$

Based on this profile function, narrower taper waist can be achieved by using a small hot-zone length in the fabrication or drawing the taper for a longer elongation distance. Tapered fiber with a short transition length can be achieved from reciprocal curve profile based on positive value of α particularly with $\alpha = 0.5$.

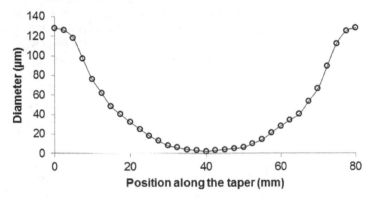

Fig. 6. A tapered fiber with decaying-exponential profile fabricated using a constant hot-zone L_0=10mm.

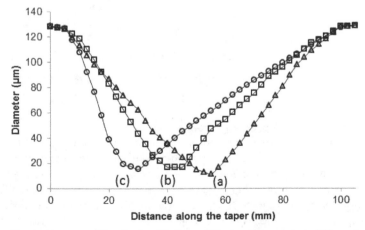

Fig. 7. Three linear taper profiles (a-c) with its smallest waist point at different positions on the tapered fibers. Profile (a) has its smallest waist point at the center of the tapered fiber.

Linear taper profile can be produced using α = -0.5 and the profile function is given by

$$r(z) = r_0 \left(1 - \frac{2z}{3L_0} \right) \tag{13}$$

Fig. 7 shows typical examples of linear taper profiles. As shown in the figure, profile (a) has the smallest waist diameter, which located at the center of the tapered fiber. By doing some simple modification on the tapering process, the smallest waist point can be shifted away from the center to one side of the tapered fiber as shown by profiles (b) and (c) in Fig. 7. These profiles are found useful in the fabrication of wideband chirped fiber bragg grating, in which the grating is written on the transition of the tapered fiber. Long linear shape tapers make good candidates for the fabrication of such devices (Frazão and et al., 2005; Mora et al., 2004; Ngo et al., 2003; Zhang et al., 2003). On the other hand, linear profile tapers can be used for optical tweezing because of its capability to converge the optical wave to a high intensity at the taper tip (Liu et al., 2006; Xu et al., 2006). Microscopic objects are attracted to the high intensity field driven by the large gradient force at the taper tip. Fig. 8 gives a good example of such tapered fiber with 15cm linear taper profile. It was produced by using a long initial hot zone length L_o = 7cm and long elongation distance.

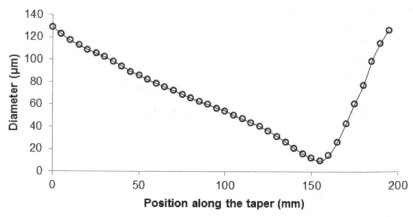

Fig. 8. The diameter of tapered fiber is linearly decreasing from ~128µm to ~10µm along the 15cm transition.

2.4 Throughput power of a degrading tapered fiber

In the high humidity environment, the concentration of water molecules in the air is high and very 'hazardous' to tapered fibers/microfiber. The increasing deposition of particles (dust) and water molecules on the microfiber is one of the major factors which causes adsorption and scattering of light that lead to perpetual decay in transmission (Ding et al., 2010). In an unprotected environment, freestanding microfibers may sway in the air due to the air turbulence. A small mechanical strength induced can cause cracks in the glass structure which may result to an unrecoverable loss in the microfibers (Brambilla et al., 2006).

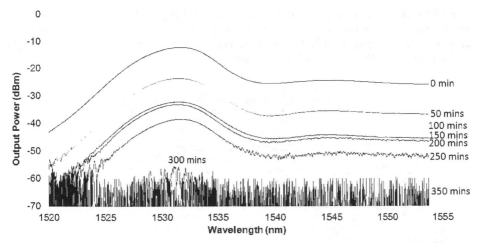

Fig. 9. The throughput power of the 10 cm long and ~3um diameter tapered fiber degrades over time.

Fig. 9 shows the output spectra of a tapered fiber in an unprotected environment. After the tapered fiber was drawn, it was left hanging in an open air and transmission spectrum was scanned and recorded every 50mins as presented in Fig. 9. Over the time, the deposition of dust and water molecules on the taper waist accumulated and the insertion loss of the tapered fiber increased over time. The output power of the tapered fiber dropped monotonically and eventually the power has gone too low beyond detection after 350mins. The throughput of the tapered fiber can be recovered by flame-brushing again as suggested in (Brambilla et al., 2006) but there is a risk that the tapered fiber can be broken after several times of flame-brushing and this solution is not practical. Despite the fact that the experiment was carried out in an air-conditioned lab where the humidity was lower (40-60%) but it was still too high for the tapered fibers. Besides, free standing tapered fibers are vulnerable to air turbulence or any sharp objects. New strategies for handling these tapered fibers are crucial for the ensuing research and fabrication of microfiber devices. In order to achieve that, this research team has been motivated to devise a packaging method to address all the problems mentioned earlier which will be discussed in the next section.

3. Packaging of microfiber

3.1 Embedding microfiber photonic devices in the low-index material
Besides the fast aging of bared microfiber in the air, the portability is another issue encountered when the microfiber is required at a different location. Moving the fabrication rig to the desired location is one way of solving the problem but it is not practical. Without a proper technique, it is risky to remove the tapered fiber from the fiber tapering rig and deliver it intact to another location. Xu and Brambilla (2007) proposed a packaging technique by embedding microfiber coil resonators in a low-index material named Teflon. Microfiber or microfiber device can be coated with or embedded in Teflon by applying some Teflon resin in solution on them and leave the solution to dry for several ten minutes. The resin is solidified after the solvent has finished evaporating from the solution, the optical properties and mechanical properties of the microfiber devices can be well preserved in the

material for a very long time (Xu and Brambilla, 2007). Jung et al. (2010) had taken slightly different approach by embedding microfiber devices in a low-index UV-curable resin. The resin is solidified by curing it with UV-light. Here, the detail procedure of embedding a microfiber device in the low-index UV-curable resin is demonstrated.

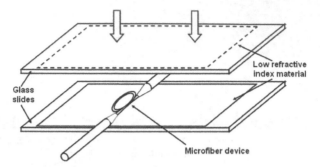

Fig. 10. Illustration of microfiber device embedded in a low refractive index material and sandwiched between two glass plates.

First, the assembled microfiber device is laid on an earlier prepared glass plate with a thin and flat layer of low refractive index material (UV-Opti-clad 1.36RCM from OPTEM Inc.) as shown in Fig. 10. The material has a refractive index of 1.36 at 1550nm. The thickness of the low refractive index material is approximately 0.5mm which is thick enough to prevent leakage of optical power from the microfiber to the glass plate. Some uncured resin is also applied on surrounding the microfiber device before it is sandwiched by another glass plate with the same low refractive index resin layer from the top. It is essentially important to ensure that minimum air bubbles and impurity are trapped around the fiber area between the two plates. This is to prevent refractive index non-uniformity in the surrounding of microfiber that may introduce loss to the system. During the tapering, coiling and coating processes, we monitored both the output spectrum and the insertion loss of the device in real time using the ASE source in conjunction with the OSA. The uncured resin is solidified by the UV light exposure for 3 ~7 minutes and the optical properties of the microfiber device are stabilized. The image of the end product is as shown in Fig. 11.

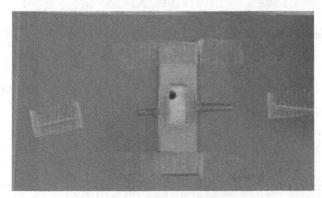

Fig. 11. The image of the end product of an embedded MLR in the low-index resin.

Fig. 12 shows recorded output spectra of the MKR at several intermediate times during the process of embedding it in a low-index UV-curable resin. The first spectrum in Fig. 12(a) was recorded right after an MKR was assembled. The fringes in the spectrum indicate that the resonance condition had been achieved in the MKR but the resonance extinction ratio remains appalling ~3dB. The MKR was benignly laid on an earlier prepared glass slide with thin layer of low refractive index material. After that, some low-index resin (refractive index ~1.36) in solution was applied onto the MKR by using a micropipette. Fig. 12(b) shows the stabilized output spectrum of the MKR and the improved resonance extinction ratio ~10dB. This phenomenon can be attributed to the reduction of index contrast; the mode field diameter (MFD) is the microfiber was expanded when it was immersed in the resin and the coupling efficiency of the MKR was altered. The changes of coupling coefficient and round-trip loss of the MKR may have induced critical coupling condition in the MKR and enhanced the resonance extinction ratio. At time = 4 minutes, UV-curing was initiated and a little bit of fluctuation is observed in the output power and extinction ratio (refer Fig. 12(c)) during the curing process. After UV-curing for 6 minutes, the output spectrum became very stable and the resin was finally solidified. Fig. 13 shows the optical microscope image of the embedded MKR in UV-curable resin.

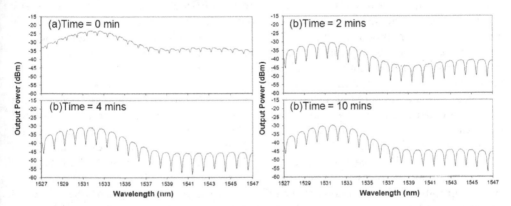

Fig. 12. Embedding an MKR in a low-index material. The time in each graph indicates when the output spectrum of the MKR is recorded. a) MKR is freestanding in the air b) some low-index resin applied on the MKR c) UV curing is initiated and d) resin is solidified.

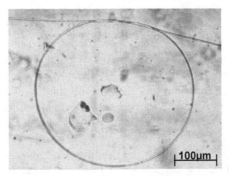

Fig. 13. Optical microscope image of an MKR embedded in UV-curable resin.

3.2 Packaging tapered fiber in a perspex case

Tapered fibers are susceptible to the air turbulence and the pollution of dust and moisture when exposed to air. It is very fragile when removed from the fiber tapering rig and maintaining the cleanliness of the tapered fiber in an unprotected condition is difficult. However, another simple packaging method had been devised to address all the difficulties mentioned. For the purpose of long term usage and ease of portability of the tapered fiber, a proper packaging process is essential. In the previous section, microfiber device is embedded in the low-index UV-curable resin to maintain the physical structure and resonance condition of the devices. Although, the refractive indices of the materials are in the range of 1.3~1.4 which is slightly lower than refractive index of silica tapered fiber and the mode can be still be confined within the tapered fiber but some optical properties such as numerical aperture (NA) and MFD will be altered due to the change in refractive index difference between silica microfiber and ambient medium when embedded in the low-index material. In the context of maintaining small confinement mode area and high optical nonlinearity, this method may not be a good idea.

In this section, a new packaging method is proposed where the tapered fiber is kept in a perspex case. The taper waist is kept straight and surrounded by the air without having any physical contact with any substance or object thus maintaining its optical properties in the air. The following part of this section provides detail descriptions of this packaging method. First, an earlier prepared perspex tapered fiber case which was made of several small perspex pieces with a thickness of 2.5mm was used in housing the tapered fiber. The perspex case mainly comprises of a lower part and upper part. Both parts of the perspex case were specially prepared in such a way that the benches at both ends of the perspex case were positioned exactly at the untapered parts of the tapered fiber. After a fresh tapered fiber was drawn, the lower part of the perspex case was carefully placed at the bottom and in parallel with the tapered fiber. That can done with the assistance of an additional translation stage. Then, the perspex case was slowly elevated upward until both benches touch both untapered parts of the tapered fiber as shown in Fig. 14(a). After that, some UV-curable optical adhesive (Norland Product, Inc) was applied to the untapered fibers that laid on the benches before the upper part of the tapered fiber case covered the tapered fiber from the top as shown in Fig. 14(b).

The UV-curable adhesive was used to adhere both the upper part and the lower part of fiber taper case. Despite that the refractive index of the optical adhesive(~1.54) is higher than silica glass (1.44) but the adhesive was only applied to untapered fiber and the light confined within the core of the fiber is unaffected. To cure the UV-curable adhesive, 9W mercury-vapour lamp that emits at ~254nm was used. Depending on the adhesive volume and its distance from the mercury-vapour lamp, the curing time takes for 2-8mins. After the adhesive was solidified and both case parts were strongly adhered to each other (Refer Fig. 14(c)). During the process illustrated in Fig. 14(a)-(c), the fiber taper was held by the two fiber holders in fiber taper rig and this helped to keep the fiber taper straight until the completion of the UV-curing process. After that the fiber taper and its case can be safely removed from the fiber holders. The fiber taper packaged inside the perspex case may remain straight permenantly. In the contrary, the fiber taper may suffer higher insertion loss if the taper fiber was bent during the packaging process. On the other hand, it is essential to prevent any physical contact between the taper waist with human hands or other objects.

The dust or moisture on the fiber taper may introduce loss to the transmission. In the final step, the perspex case was sealed by wrapping it with a piece of plastic wrap. This can minimize the pollution of dust or air moisture in the perspex case for a very long period of time. This fiber taper can be kept in storage for a week and possibly a fortnight without having an increment of loss more than 1.5dB however it is subject to taper dimension and its usage in the experiment.

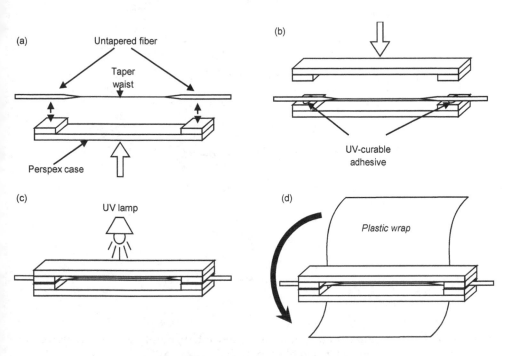

Fig. 14. Schematic illustration for tapered fiber packaging process.

To observe the characteristic of the tapered fiber as well as the reliability of the tapered fiber case over time, an observation on the transmission spectrum was conducted on the packaged tapered fiber for 6 days. Figs. 16(a) and (b) show the 6 days output spectra and output power observation, respectively for the packaged tapered fiber. Unlike the monotonic decrease in throughput power observed in Fig. 9(a), the curve of every transmission spectrum is closely overlaid to each other with a small power variation <1.2dB in the graph. Refer to Fig. 16(b), the variation of the total output power is spontaneous which can be attributed to the fluctuation of power at the ASE source and change of ambient temperature. In comparison with the taper fiber without packaging, obviously Perspex case plays its role well in preserving the tapered fiber to a longer lifespan; it enables portability and allows integration with more complex optical fiber configurations away from the fiber tapering rig.

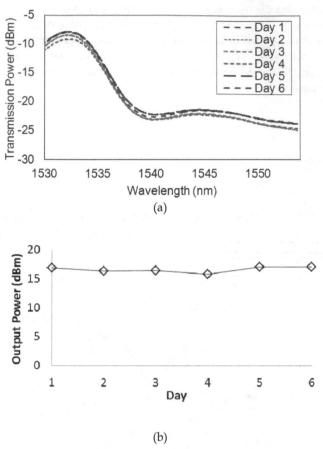

Fig. 16. The 6 days comparison of (a) output spectrum and (b) output power variation of the 10cm long and ~3um diameter tapered fiber packaged in perspex case.

4. Optical microfiber devices

Optical microfiber devices have attracted growing interest recently especially in their simple fabrication methods. This is due to a number of interesting optical properties in this device, which can be used to develop low-cost, miniaturized and all-fiber based optical devices for various applications (Guo et al., 2007). For instance, many research efforts have been focused on the development of microfiber/nanofibers based optical resonators that can serve as optical filters, which has many potential applications in optical communication, laser systems (Harun et al., 2010), and sensors (Hou et al., 2010; Sumetsky et al., 2006). Many photonic devices that are conventionally fabricated into lithographic planar waveguides can also be assembled from microfibers. Recently, there are many microfiber devices have been reported such as MLR (Harun et al., 2010; Sumetsky et al., 2005), MCR (Sumetsky, 2008; Sumetsky et al., 2010; Xu and Brambilla, 2007; Xu et al., 2007), MKR (Jiang et al., 2006; Lim et al., 2011; Wu et al., 2009), reef knot microfiber resonator as an add/drop filter (Vienne et al., 2009) , microfiber mach-zehnder interferometer (MMZI) (Chen, 2010; Li and Tong, 2008) and etc. These microfiber based devices have the similar functionalities, characteristics and possibly the same miniaturizability with the lithographic planar waveguides. In future, these microfiber based devices may be used as building blocks for the larger and more complex photonic circuits. In this chapter, the transmission spectrum and the corresponding theoretical model of three microfiber based devices are presented, namely MLR, MKR and MMZI. In addition, some of the important optical properties of these devices will be reviewed and discussed.

4.1 Microfiber Loop Resonator (MLR)

MLRs are assembled from a single mode microfiber, which is obtained by heating and stretching a single mode fiber. In the past, many MLRs have been demonstrated. For instance and Bachus (1989) assembled a 2mm diameter MLR from an 8.5μm tapered fiber where the coupling efficiency can be compromised by the large thickness of the microfiber. However the deficiency was compensated by embedding the MLR in a silicone rubber which has lower and near to the refractive index of silica microfiber. The transmission spectrum with a frequency spectral range (FSR) of 30GHz is observed from the MLR (Caspar and Bachus, 1989). Later on, Sumetsky et al. had demonstrated the fabrication of MLR from a ~1μm diameter waist microfiber which has the highest achieved loaded Q-factor as high as 120,000 (Sumetsky et al., 2006). Guo et al. demonstrated wrapping a ~2 μm diameter microfiber loop around copper wire which is a high-loss optical medium. By manipulating the input-output fiber cross angle, the loss induced and the coupling parameter in the resonator can be varied. In the condition when the coupling ratio is equivalent to the round-trip attenuation, the MLR has achieved critical coupling and the transmission of resonance wavelength is minimum. In their work, critical coupling condition have been achieved which resonance extinction ratio as high as 30dB had been demonstrated (Guo et al., 2007; Guo and Tong, 2008).

4.1.1 Fabrication of MLR

Fig. 19 shows an example of ~3mm loop diameter MLR assembled from a ~2.0 μm waist diameter microfiber. Similar to other optical ring resonators, MLR has a 'ring' but manufactured from a single mode microfiber. This fabrication can be carried out with the assistance of two 3D translation stages as illustrated in Fig. 20. By aligning the three–axial position of each translation stage and twisting one of the pigtails, the microfiber is coiled into a loop. If the microfiber is sufficiently thin, the van der Waals attraction force between two adjacent microfibers is strong enough to withstand the elastic force from the bending microfiber and maintain the microfiber loop structure. The diameter of the loop can then be reduced by slowly pulling the two SMFs apart using the translation stages. Due the large evanescent field of the microfiber, a coupling region is established at the close contact between the two microfibers and a closed optical path is formed within the microfiber loop. Since the MLR is manufactured from an adiabatically stretched tapered fiber, it has smaller connection loss because microfiber based devices do not have the input-output coupling issue encountered in many lithographic planar waveguides. Despite the difference in the physical structure and fabrication technique between MLR and the conventional optical waveguide ring resonator, they share the same optical characteristics.

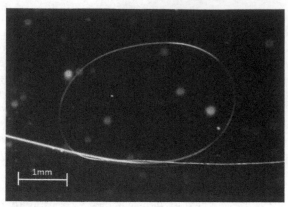

Fig. 19. Optical microscope image of an MLR.

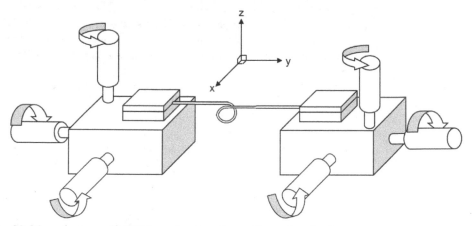

Fig. 20. Manufacture of MLR by using two three-dimensional stages.

4.1.2 Theory

The microfiber guides light as a single mode waveguide, with the evanescent field extending outside the microfiber. This evanescent field depends on the wavelength of operation, the diameter of the fiber and the surrounding medium. If the microfiber is coiled onto itself, the modes in the two different sections can overlap and couple to create a resonator. On every round-trip of light in the loop, there are fractions of light energy exchange between the two adjacent microfibers at the coupling region, the input light is allowed to oscillate in the closed loop and the resonance is strongest when a positive interference condition is fulfilled which can be related to this equation.

$$n\lambda_R = L \qquad\qquad (14)$$

where L is the round-trip length, λ_R is the wavelength of the circulating waves and n is an integer. Positive interference occurs to those circulating waves and the wave intensity is building up within the microfiber loop. The relationship in Eqn (14) indicates that each wavelength is uniformly spaced and periodic in frequency, a well-known characteristic of an optical multichannel filter. The amplitude transfer function for MLR is given as (Sumetsky et al., 2006);

$$T = \frac{\exp(-\alpha L / 2)\exp(j\beta L) - \sin K}{1 - \exp(-\alpha L / 2)\exp(j\beta L)\sin K} \quad (15)$$

$\sin K$ denotes the coupling parameter where $K = \kappa l$, κ is the coupling coefficient and l is the coupling length. For every oscillation in the MLR, the circulating wave is experiencing some attenuation in intensity attributed to non-uniformity in microfiber diameter, material loss, impurity in the ambient of microfiber and bending loss along the microfiber loop. However, these losses can be combined and represented by a round-trip attenuation factor, $\exp(\alpha L/2)$ in Eqn (15) while $\exp(j\beta L)$ represents phase increment in a single round-trip in the resonator. The intensity transfer function is obtained by taking the magnitude squared of the amplitude transfer function in Eqn (15).

$$|T|^2 = \frac{\exp(-\alpha L) + \sin^2(K) - 2\exp(-\alpha L / 2)\sin(K)\cos(\beta L)}{1 + \exp(-\alpha L)\sin^2(K) - 2\exp(-\alpha L / 2)\sin(K)\cos(\beta L)} \quad (16)$$

The resonance condition occurs when

$$\beta L = 2m\pi \quad (17)$$

where m is any integer. The critical coupling occurs when

$$\sin K_c = \exp(-\alpha L / 2) \quad (18)$$

The FSR is defined as the spacing between two adjacent resonance wavelengths in the transmission spectrum which is given by

$$\text{FSR, } \Delta\lambda \approx \frac{\lambda^2}{n_{eff}L} \quad (19)$$

or

$$\Delta\lambda \approx \frac{\lambda^2}{n_{eff}\pi D} \quad (20)$$

where D is the diameter of the circular loop.
In addition to the characteristic parameters mentioned earlier, Q-factor and finesse F are two important parameters that define the performance of the MLR. The Q-factor is defined as the ratio of resonance wavelength to the bandwidth of the resonance wavelength, the full wave at half maximum, FWHM (Refer Fig. 21). It is given as;

$$Q = \frac{\lambda}{FWHM} \quad (21)$$

The finesse is defined as the FSR of the resonator divided by the FWHM;

$$F = \frac{FSR}{FWHM} \qquad (22)$$

Due to the narrow bandwidth at the resonance wavelengths, MLR also functions as a notch filter (Schwelb, 2004). The attenuation at the resonance wavelength can be used to filter out/drop the signal from specific channels in the WDM network by suppressing the signal power. In DWDM network, the spacing between two adjacent channels in the network is small therefore notch filter with narrow resonant bandwidth is preferable so that the signals from adjacent channels are unaffected by the attenuation in the drop channel. Based on the relationship in Eqn (22), narrow resonant bandwidth (FWHM) can be found in high finesse filter.

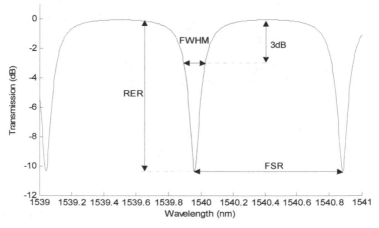

Fig. 21. Typical transmission spectrum of an MLR. The labels in the graph indicates the terminology used in the chapter. RER is an abbreviation for resonance extinction ratio.

4.1.3 Transmission spectra of MLRs

The typical transmission spectra of an MLR with different FSRs are shown in Fig. 22. For better clarity of viewing, the transmission spectra with different FSRs are presented in a increasing order from the top to the the bottom in the figure. These transmission spectra were recorded from a freestanding MLR in the air, started from a large loop diameter and the diameter is decreasing in step when the two microfiber arms of the MLR are stretched. Exploiting the van der Waals attraction force between the two microfibers in the coupling region, the resonance condition of the MLR can still be maintained during the stretching of microfiber. In the measurement, the loop diameters are at approximately 1.9mm, 1.4mm, 1.1mm, 0.8mm and 0.6mm which corresponds to FSR values of 0.275nm, 0.373nm, 0.493nm, 0.688nm and 0.925nm, respectively in the C-band region as shown in Fig. 22. These variations of FSR and loop diameter are very consistent with the reciprocal relationship expressed in Eqn (23). The loop diameter of an MLR is restricted by the microfiber elastic force, the smaller is the loop diameter the greater is the elastic force. Thus, it is difficult to keep the microfiber loop in shape when the loop diameter is very small and the MLR loses its resonance condition when the loop opens.

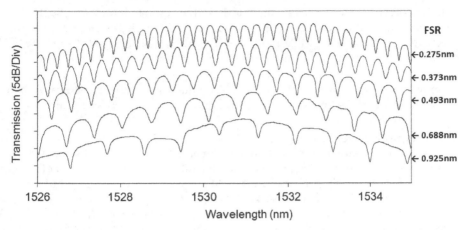

Fig. 22. Transmission Spectra of an MLR with increasing FSR (from top to bottom).

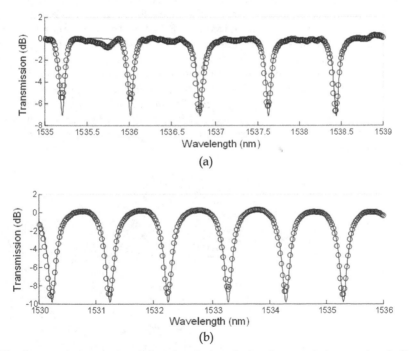

Fig. 23. The fitting of experimental data (circles) with the characteristic equation (solid line). (a) Q-factor ~18,000 and finesse ~9.5 (b) Q-factor ~5700 and finesse ~3.8

Fig. 23(a) and Fig. 23(b) show the fitting of experimental data with the (intensity) analytical model based on characteristic equation in Eqn (19). The best-fit parameters for the transmission spectrum in Fig. 23(a) are $L = 2.03$mm, exp $(-aL/2) = 0.8853$ and $\sin K = 0.7354$. The measured FSR ~0.805nm from transmission spectrum in Fig. 23(a) is in agreement with the calculated FSR ~ 0.807nm. The bandwidth at the resonance wavelength, FWHM is ~0.085nm indicates that the Q-factor and finesse of the MLR are 18,000 and ~9.5. The best fit parameter for Fig. 23(b) are L = 1.61mm, $\sin K$ = 0.5133 and $\exp(-aL/2)$ = 0.6961. The measured FSR and FWHM are ~1.02nm and ~0.27nm respectively which indicate the values of Q-factor and finesse are ~5700 and ~3.8. In the comparison between the two spectra, the finesse provides a good representation in weighting the bandwidth between passband and stopband. The higher is the finesse the narrower is the stop-band compared with pass-band.

4.2 Microfiber Knot Resonator (MKR)

MKR is assembled by cutting a long and uniform tapered fiber into two. One tapered fiber is used for the fabrication of microfiber knot while the other one is used to collect the output power of the MKR by coupling the two tapered fiber ends and guides the output light back to an SMF. The fabrication of microfiber knot can be done by using tweezers. The coupling region of the MKR is enclosed by a dashed box in Fig. 24 where the two microfibers intertwisted and overlapped in the resonator. In comparison with MLR, MKR does not rely on van der Waals attraction force to maintain the coupling region yet it can achieve stronger coupling due to the rigid intertwisted microfibers structure at the coupling region. The knot structure can withstand strong elastic force of the microfiber and maintain a rigid resonator structure with a more stable resonance condition. Based on the same microfiber diameter, MKR of smaller knot diameter can be easily manufactured than that of MLR. However, MKR suffers a setback in a high insertion loss due to the cut-coil-couple process where the evanescent coupling between output microfiber and collector microfiber contributes a large fraction in the total insertion loss. The microfiber diameter in the range of 1~3μm is preferable because thinner microfiber is very fragile and it breaks easily in the fabrication of MKRs. Nonetheless, the operating principle of MKR is identical to MLR as it is based on self-touching configuration thus the same characteristic equation can be used to describe the transmission spectrum of MKR.

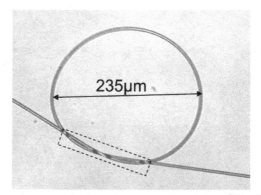

Fig. 24. Optical microscope image of an MKR.

4.2.1 Transmission spectra of MKRs

MKR offers better capability in achieving smaller knot diameter. The knot can withstand the strong elastic force of microfiber and it can achieve a small knot diameter that cannot be achieved in the MLR. Fig. 25 shows transmission spectra of an MKR assembled in the laboratory. The transmission spectra are presented in an increasing order from the top to bottom of the figure and the values are 0.803nm, 1.030nm, 1.383nm, 1.693nm, 2.163nm and 2.660nm within the vicinity of 1530nm. The corresponding knot diameters are approximately 640μm, 500μm, 370μm, 310μm, 240μm and 190μm.

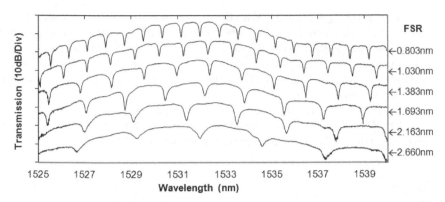

Fig. 25. Transmission Spectra of an MKR with increasing FSR (from top to bottom).

4.2.2 Resonance condition of microfiber knot resonator immersed in liquids

Recently, microfiber resonators are suggested in numerous applications particularly in the sensing applications (Lim et al., 2011; Sumetsky et al., 2006). The operating principles of these sensors rely on the characteristics of the resonance, the variation of the position of resonance wavelength and the resonance extinction ratio with the sensing parameters, temperature, refractive index and etc. (Wu et al., 2009; Xu et al., 2008). The resonance condition of a resonator relies on the index contrast between microfiber and its ambient medium, evanescent field strength and distance between two microfibers in the coupling region. Large evanescent field which can be found in thinner microfibers is one of the solutions to achieving higher coupling in microfiber resonators. The large fraction of light intensity in the evanescent field allows stronger mode interaction between two microfibers and yields high coupling coefficient. Caspar et al. suggest embedding the microfiber resonator into a medium that has a slightly lower refractive index than that of silica. Due to the small index contrast, the microfiber has a larger evanescent field which yields stronger coupling in the resonator (Caspar and Bachus, 1989; Xu and Brambilla, 2007). Besides being used as a post-fabrication remedy for improving the resonance condition of the resonator, embedding also offers good protection from the fast aging process and enabling portability for the microfiber devices. Vienne at el. have reported that when a microfiber resonator is embedded in low-index polymer, the optimal resonance wavelength is down-shifted by ~20% (Vienne et al., 2007). However, there is very few literatures that provide mathematically analysis on the effect of embedding in low index contrast medium to the resonance condition of the resonator. In order to achieve a better understanding, an experiment on an MKR immersed in liquid solutions was carried out. MKR was used in the experiment due to its rigid knot structure and strong interfiber coupling. The knot structure and resonance condition could be easily maintained during the immersing process.

Unlike MLR that exploits van der Waals attractive force to maintain the structure of the loop, MKR has a more rigid knot structure with interfiber twisted coupling. Nonetheless, both MLR and MKR share the same optical properties, the same transmission equation can be used to describe both structures. Fabrication of an MKR started with fiber tapering using heat and pull technique. After a single-mode biconical tapered fiber was drawn, it was cut at one third part of the waist which the longer section of tapered fiber was used for the fabrication of knot by using tweezers. Then the second section was used as collector fiber by evanescent coupling (Tong et al., 2003) with the output port of the MKR. Immediately after that, the transmission spectrum of the freestanding MKR in the air was recorded by an OSA. After that, the MKR was embedded in propan-2-ol solution that has a refractive index (RI) of 1.37. First, the MKR was slowly laid horizontally on an earlier prepared flat platform deposited with a thin layer of propan-2-ol. Using a micropipette, a small volume of propan-2-ol solution was dropped onto the MKR and had it entirely immersed in the solution. The structure of the microfiber knot was intact and the resonance was maintained. This is the crucial part that distinguishes MKR from MLR. It is very difficult to maintain the loop structure and resonance of MLR when immersed in the liquid.

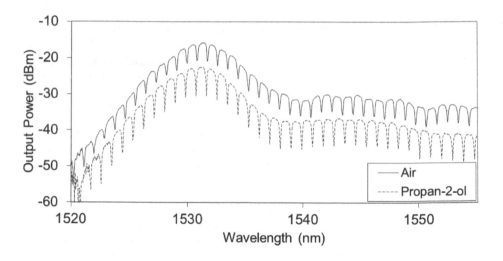

Fig. 26. The transmission spectra of MKR in the air (solid) and propan-2-ol solution (dashed).

Fig. 26 shows the overlaid transmission spectra of the MKR in the air (solid) and solution (dashed). Refering to the peak powers of both spectra, it is easy to determine that the MKR had suffered an additional ~7dB excess loss after it was immersed in the solution. The drop in coupling efficiency at the evanescent coupling between MKR output microfiber and collector microfiber constituted a large fraction in this excess loss. On the other hand, the resonance extinction ratio of the MKR had improved from ~5dB to ~8dB. In the analysis of resonance characteristics, the coupling parameter, $\sin\kappa\ell$ and round-trip attenuation factor of MKR, $\exp(-aL/2)$ can be extracted from the best-fit curves (lines) for the offset experimental data (circles) in Fig. 27(a) and Fig. 27(b) based on the transfer function in (18).

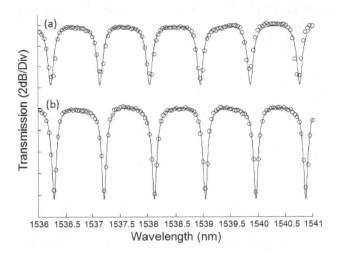

Fig. 27. The offset experimental data (circles) with its best fit curve (solid line) (a) air, RI ~1.00 (b) propan-2-ol, RI ~1.37.

In the air, the best fit parameters for the transmission spectrum in Fig. 27(a) are $\sin\kappa\ell$ = 0.6207 and $\exp(-aL/2)$= 0.8547. When the MKR was immersed in the propan-2-ol solution, the best fit parameters for Fig. 27(b) are $\sin\kappa\ell$ = 0.6762 and $\exp(-aL/2)$ = 0.8361. The reduction in round-trip attenuation factor can be attributed to the small index contrast between microfiber and ambient medium when immersed in the solution so that the bending loss at the microfiber knot is higher. The output - collector coupling loss is excluded from this analysis as it only affects the total output power (position in the vertical axis) and it can be eliminated in the offset spectrum. Based on Eqn (18), the resonance state of the resonator can be determined from the following expression

$$\delta = \left| \frac{\exp(-\alpha L/2) - \sin\kappa l}{1 - \exp(-\alpha L/2)\sin\kappa l} \right| \qquad (23)$$

Smaller value of δ indicates that the state of resonance is closer to the critical coupling condition and it yields larger resonance extinction ratio. In fact, the resonance extinction ratio can be estimated by

$$RER \sim 20\log_{10}\delta \qquad (24)$$

Comparing the two spectra, the spectrum in Fig. 27(b) has higher coupling and smaller round-trip attenuation factor which give smaller value in δ = 0.3679 if compared with δ = 0.5060 obtained from the spectr $\sin\kappa\ell$ and resulting the lower coupling value. The next experimental data may provide an example for such self-defeating scenario.The transmission spectra of an MKR in the air and low-index UV-curable resin (UV-Opti-clad 1.36RCM from OPTEM Inc.) with an RI of ~1.36 are as shown in Fig. 28(a) and Fig. 28(b) respectively. The coupling parameter, $\sin\kappa\ell$ had dropped from 0.7132 to 0.6247 when it was immersed in the water. On the other hand, the round-trip attenuation factor, $\exp(-aL/2)$ suffers greater fall from 0.9432 to 0.7538. In spite of that, the resonance extinction ratio had increased from ~2dB to ~10dB. This is in agreement with the decreasing value of δ from 0.7027 to 0.2440 and the state of resonance is closer to critical coupling condition.

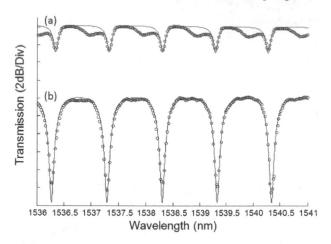

Fig. 28. The transmission spectra of MKR in different ambient mediums (a) air, RI ~1.00 (b) low-index resin, RI ~1.36.

Immersing MKR in a near-index medium do not always promise an improvement in the resonance condition or RER. There is a possibility that the changes in round-trip attenuation factor and coupling parameter yield larger value of δ and decreases the RER. Fig. 29 provides an example for this scenario. The best-fit parameter for the experimental data in the air (solid) are $\sin\kappa\ell = 0.6235$ and $\exp(-aL/2) = 0.8145$ respectively. After the MKR is immersed in the water (dashed), the values have varied to $\sin\kappa\ell = 0.7833$ and $\exp(-aL/2) = 0.9339$. In the air, the low value of round-trip attenuation factor can be attributed to the large amount of deposited dust on the microfiber surface which was introduced from the tweezers during the fabrication of microfiber knot. After it was immersed in the water, some portion of the dust might have been 'washed' away and that increases the round-trip attenuation factor. The value of δ has decreased from 0.3881 to 0.5609 which is an indication of the resonance state deviates from critical coupling.

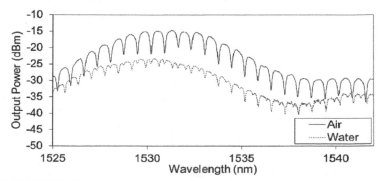

Fig. 29. Example of an MKR with decreased resonance extinction ratio after it is immersed in the water (RI ~1.33).

The purpose of using liquid solutions of different RIs is to investigate the influence of different index-contrasts to the resonance characteristics of the MKR. However, there was no significant indication observed in the experiment showing difference between the solutions. The only obvious changes were observed at the moment when the MKRs were immersed in the solutions. It is believed that the index contrasts induced by these three liquids are within a narrow range from 0.07 to 0.11 and the differences among them are too small to make significant impact on the characteristics of the MKR. On the other hand, we believe that that the microfiber waist diameter and orientation of the microfibers in the coupling region have an important relationship with the resonance of MKR. More investigations pertaining to those parameters are needed.

4.2.3 Polarization dependent characteristic

Microfiber based resonator exhibits strong dependence on its input polarization state. Similar characteristic was reported by Caspar et al. (Caspar and Bachus, 1989) where the extinction ratio of the resonator varies with the change of input state of polarization.

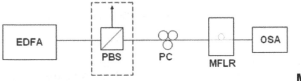

Fig. 30. Experimental set-up to investigate the polarization dependent characteristic of the MLR (Lim et al., 2011). Polarized wideband source from EDFA is acquired with the aid of PBS (Dashed box), an unpolarized wideband source can be obtained by removing PBS from the setup.

In the investigation of polarization state dependent characteristics, a simple experimental setup as shown in the Fig. 30 is established. First, the unpolarized ASE source from an EDFA was linearly polarized by a PBS, followed by a PC for controlling the state of polarization (SOP) of the polarized source before it was fed into the MLR. Fig. 31 shows the transmission spectra of the MKR at various resonance conditions of the MLR, which was obtained at different input wave SOP. Both spectra (i) in Fig. 31(a) and Fig. 31(b) show the transmission of the MLR based on an unpolarized input wave (without PBS), which resonance condition was unaffected by the adjustment of the PC. In contrast, the resonance condition for the MLR with polarized input wave was sensitive to the PC adjustment. By carefully adjusting the PC, the resonance extinction ratio could be enhanced or reduced as depicted in spectra (ii) and (iii) of both Fig. 31(a) and Fig. 31(b). Nevertheless, the wavelength of each peak and the FSR remained unchanged regardless of input wave SOP. It is appropriate to attribute this phenomenon to the polarization dependent coupling in the MLR where the interfiber coupling, twisting and alignment of microfiber in the coupling region are accounted for the coupling coefficient difference between two orthogonal polarization states (Bricheno and Baker, 1985; Chen and Burns, 1982; Yang and Chang, 1998). Associated with the coupling coefficient, the resonance extinction ratio of a microfiber based resonator can be improved by using an optimized polarized input light. However, more efforts are required for more in-depth study and to explore the possible applications in many fields such as multi-wavelength laser generation and sensing.

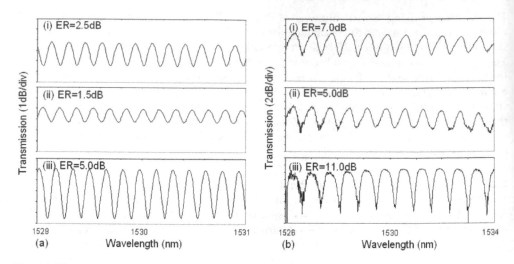

Fig. 31. The spectra of two MLRs for different input wave SOP. (a) FSR=0.162nm at 1530nm (b) FSR=0.71nm at 1530nm.

4.2.4 Thermal dependent characteristic

Microfiber based devices are very sensitive to variation of ambient temperature due to the strong dependence of the microfiber dimension and refractive index on temperature. In a thermally unstable environment, these devices may experience thermal drift in the transmission spectrum and fluctuation in the transmission power. However, this problem can be alleviated by the placing the devices in a temperature controlled housing (Dong et al., 2005). Sumetsky et al. demonstrated a MLR based ultrafast sensor for measurement of gas temperature. Taking advantage of the close contact between the MLR and air, the change in gas temperature in the ambient of MLR can be determined from the transmission power at the resonance wavelengths within a short response time of several microseconds (Sumetsky et al., 2006). Besides, the positions of resonance wavelengths are found to be sensitive to temperature change. The spectral shift of an MLR can be expressed in a linear function of temperature. The property enables temperature measurement based resonance wavelength shift with higher accuracy (Wu et al., 2009; Zeng et al., 2009).

MKRs exhibit the similar optical properties with MLRs. The free spectral range of MKR takes in the form of Eqn (22). Based on this equation, the variations in effective index n_{eff} and round-trip length L may lead to transmission spectral shift and their relationship can be expressed as

$$\frac{\Delta \lambda_{res}}{\lambda_{res}} = \left(\frac{\Delta n_{eff}}{n_{eff}} + \frac{\Delta L}{L} \right)_{Temp.} \tag{25}$$

In relation with temperature, both terms on the right hand side of Eqn (25) expresses two linear thermal coefficients; thermal-optic coefficient (TOC) and thermal expansion coefficient (TEC) [13]. With this interpretation, Eqn (25) can be rewritten as

$$\frac{\Delta\lambda_{res}}{\lambda_{res}} = (\alpha_{TOC} + \alpha_{TEC})\Delta T \tag{26}$$

Fig. 32(a) shows the transmission spectra of MKR at temperatures of 30°C, 35°C and 40°C. The spectral shift is approximately 26pm for every temperature increment of 5°C and the linearity between wavelength shift and temperature change can be seen from the linear fitting of experimental data in Fig. 32(b).

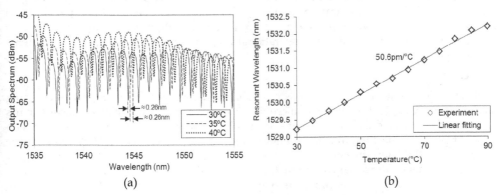

(a)   (b)

Fig. 32. (a) The output spectra of an MKR at temperature of 30°C(solid), 35°C(dashed) and 40°C (dotted) (b) The temperature response of the MKR has spectral sensitivity of 50.6pm/°C.

This characteristic has opened up new possibilities for temperature sensing and spectral control based on temperature manipulation. It provides a solution for stabilizing the spectrum of the device which often affected by the thermal drift. Dynamic spectral shift in optical filter can be realized by exploiting this characteristic. In addition, the insusceptibility of fiber-optic components to electrical noise has made these devices very attractive for many industrial sensing applications.

4.2.5 Microfiber knot resonator based current sensor

A variety of fiber optic based current sensors have been investigated in recent years using mainly a single mode fiber (SMF) of clad silica. They are typically divided into two categories, where one is based on Faraday Effect and the other is based on thermal effect. The former is capable to remotely measure electrical currents, but the device requires a long fiber due to the extremely small Verdet constant of silica. The latter needs a short length of fiber but requires complex manufacturing techniques to coat fibers with the metals. Recently, a resonant wavelength of the MLR has been experimentally reported to shift with electric current applied to the loop through a copper rod. An acceptable transmission loss is achieved despite the fact that copper is not a good low-index material to support the operation of such structure (Guo et al., 2007). This finding has opened up a way to enable dynamic and efficient spectrum control for optical filters by manipulating an electric current dependence spectral shift characteristic of the microfiber based resonator. Microfiber based devices have a strong dependence on temperature due to the thermal expansion characteristic and thermo-optic effect of silica glass. As discussed earlier in Section 12, the transmission spectrum of a microfiber based device shifts as the ambient temperature varies and the relationship between these two variables is well described by the linear equation in Eqn. (26).

In this section, spectral tunable MKR is demonstrated based on the idea of thermally induced resonant wavelength shift. By manipulating the applying electric current through the microfiber knot wrapped copper wire, the copper wire acts as a heating element and induces temperature change in the MKR. The transmission spectrum of the MKR shifts corresponds to the temperature change. These modified MKRs can be used as low-cost and fast response tunable optical filters which are useful in the applications of optical signal processing, WDM communication and etc. On the other hand, this opto-electrical configuration may operate as a dynamic current sensor with strong immunity to electric noise. In addition, it has a dynamic operational range extending to the regime of extreme high temperature or pressure.

Fabrication

First, a ~2μm diameter silica microfiber is fabricated from a SMF using flame-brushing method (Graf et al., 2009). Then the microfiber is cut and separated into two unequal parts in which the longer one is used in the knot fabrication and the other one is used as a collector fiber to collect the transmitted light from the MKR (Jiang et al., 2006). During the fabrication of the knot, the copper wire is inserted into the knot which diameter is bigger than the diameter of the copper wire (Refer Fig. 33(a)). The light path from the knot resonator is completed by coupling the two microfiber ends. At least ~3 mm of coupling length between two microfibers is required to achieve strong van der Waal attraction force to keep them attached together. The microfiber knot diameter is then reduced and fastened on the copper wire by pulling microfibers from both arms of the microfiber knot as illustrated in Fig. 33(b).

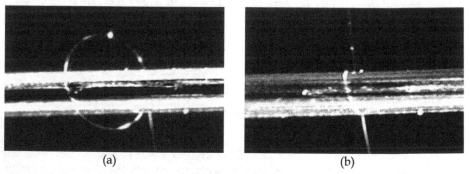

(a)   (b)

Fig. 33. Optical microscope image of MKR tied on a copper wire.

The optical characteristics of the resonator are strongly affected by the tensile strain on the microfiber arms of the MKR induced by the pulling on the microfibers arms. It is essential to reduce the tension on the both arms of the MKR by moving the fiber holders a bit closer to the microfiber knot after the knot is fastened. In spite of that, there is a very little change at the knot diameter and the resonance condition of the MKR remains good and stable after the tension is released.

Theoretical Background

By wrapping microfiber on a current loaded conductor rod, the conductor rod acts as a heating element. It generates heat and increases the temperature of the microfiber. As discussed earlier in Section 13, the transmission spectrum of a microfiber based device shifts as its ambient temperature varies and the relationship between these two variables is well described by the linear equation in Eqn. (26). Consider the linear relationship between the temperature change and heat energy generated by the conducting current, the relationship between wavelength shift and the conducting current, I can be expressed in the form of

$$\frac{\Delta \lambda_{res}}{\lambda_{res}} \propto \frac{\rho I^2}{A} \tag{27}$$

where ρ and A represent the conductor resistivity and the cross sectional area of the conductor rod, respectively. The term ρ / A in Eqn. (27) is equivalent to the resistance per unit length of the conductor material. The resistivity of the copper rod is 1.68×10^{-8} Ω m.

Current Response

For optical characterization of the MKR, broadband source from amplified spontaneous emission is first launched into and guide along the SMF and then squeezed into the microfiber through the taper area. The light transmitted out from the MKR is collected by the collection fiber and measured by an OSA. The optical resonance is generated when light traversing the MKR. When an alternating current flows through the copper wire, heat is produced in the wire to change temperature. Because the MKR is in contact with the copper wire, any temperature changes will influence the refractive index and the optical path length of the MKR.

Fig. 34 shows the resonant spectral of the MKR tied on a copper wire with various current loadings. In the experiment, the applying current is uniformly increased from 0 to 2A.

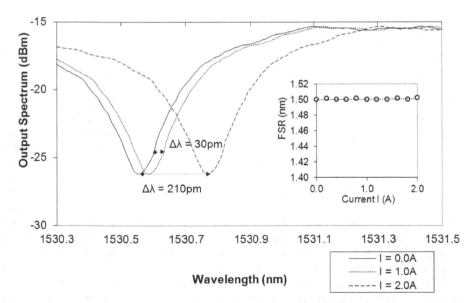

Fig. 34. Resonant wavelength shift of the MKR tied on a copper rod loaded with different current. Inset shows unchanged FSR with the increasing current.

In the spectrum, the resonant wavelength shifts to a longer wavelength the increasing of conducting current in the copper wire. The response time of the wavelength shift is approximately 3s and the spectrum comes to steady condition after 8s. Therefore, each spectrum is recorded at ~10s after the copper wire is loaded with an electric current. At loading current I = 1.0A, the resonant wavelength is shifted by ~30pm from 1530.56nm to 1530.59nm and at I = 2.0A, the resonant wavelength is further shifted to 1530.77nm, 210pm from the original wavelength. Inset of Fig. 34 shows the free spectral range of the transmitted spectrum against the applying currents. As shown in the inset, FSR of the MKR remains unchanged at 1.5 nm with the increasing current. The calculated Q factor and finesse of the MKR are ~4400 and 4.3 respectively. It is also observed that the transmission spectrum always shift towards the longer wavelength direction with increasing current regardless of the current flow direction, and the spectrum returns to original state once current supply is terminated.

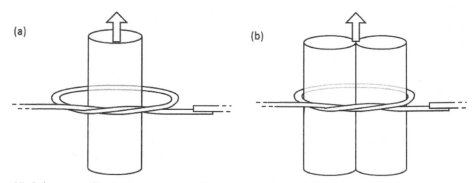

Fig. 35. Schematic illustrations of microfiber knot tied on (a) single copper wire (b) two copper wires with identical wire diameter of ~200µm.

Fig. 35(a) and Fig. 35(b) give schematic illustrations of MKR wrapped on a single copper rod and two copper rods respectively. The measured FSR and knot diameter of the single-rod MKR are 1.7nm and ~317µm respectively while for the two-rod MKR, the measured FSR and knot diameter are 1.46nm and ~370µm. An direct current is applied through the copper wire and the resonant wavelength shift is investigated against the applying current. At small current of < 0.5A, no significant resonant wavelength shift is observed. Beginning at 0.6A, the resonant wavelength shifts gradually toward the longer wavelength. At applying current of 2.0A, a wavelength shift of 0.208nm and 0.09nm are achieved with single and two copper wires configurations of Fig. 35(a) and Fig. 35(b), respectively. Fig. 36 shows the resonant wavelength shift against a square of current (I^2) for both configurations. The data set of each configuration can be well fit with a linear regression line with a correlation coefficient value r > 0.95. This justifies the linear relationship stated in Eqn. (27). In comparing the conductor wire cross-sectional area between the two configurations, the two-wire configuration is twice larger than the single-wire configuration. Based on the relation in Eqn. (27), the tuning slope of the wavelength shift with I^2 of the two-wire configuration should be a half of the single-wire configuration. The slope of each linear line is 51.3pm/A^2 (single rod) and 19.5pm/A^2 (two rods) nonetheless it is reasonable to attribute the mismatch between the analysis and experiment to the different orientation and position of the rod(s) in the MKR. The tuning slope of the current sensor can be further increased by using different conductors with higher resistivity such as nichrome, constantan, graphite and etc which are commonly used as heating elements. However, the suitability in integration with microfiber or other opto-dielectric device remains uncertain.

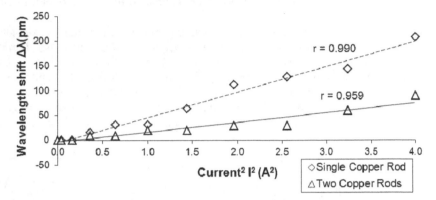

Fig. 36. Current response of MKRs based on single-wire and two-wire configurations. The calculated resistance of the single copper wire and two copper wire are 0.53 $\Omega \cdot m^{-1}$ and 0.26$\Omega \cdot m^{-1}$ respectively.

4.3. Microfiber Coil Resonator (MCR)

Microfiber coil resonators (MCRs) possess similar functionality with other microfiber resonators. It is useful for the applications of optical filtering, lasers and sensors. Additionally, it can be employed as an optical delay line for the optical communication network with small compactness. It is fabricated by winding a long microfiber on a low-index dielectric rod or a rod coated with low-index material (Sumetsky et al., 2010). The helical structure of microfiber coil enables propagation of light along the microfiber, across between the turns of microfiber in the forward and backward directions as illustrated in Fig. 37.

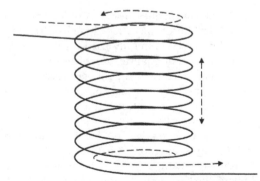

Fig. 37. Helical structure of an MCR and the propagation direction of the light in the resonator.

Fig. 38 shows the transmission spectra of an MCR of increasing number of microfiber turns. The microfiber was wound on a 0.5mm-diameter low-index coated rod. Basically, the optical characteristics a 1-turn MCR is exactly identical to that of MLR, for instance the interference fringes in the MCR transmission spectrum are equally spaced as shown in Fig. 38(a). From the spectrum, the measured FSR is ~0.8nm and the estimated diameter of the coil is ~0.6mm which is slightly larger than the rod diameter. When making additional turns to the coil, it is important to ensure overlapping or touching between turns to establish coupling between them. For every additional turn is made, the transmission spectrum of the MCR is altered. Figs. 38(b)-(d) show the transmission spectra of a 2-turn, 3-turn and 4-turn MCR fabricated in the laboratory. Consider the elastic force from the bent microfiber; it is very difficult to maintain the resonance condition and increasing the number of turns at the same. With the assistance of microscope, the ensuing coiling work can be alleviated. Nonetheless, the reproducibility of MCR was difficult and tedious. Compared with the other microfiber resonators, MCRs have more complicated light propagation properties (Hsu and Huang, 2005).

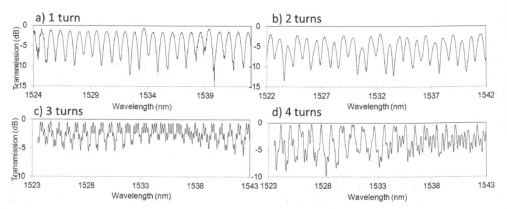

Fig. 38. Transmission spectra of a) 1 turn b) 2 turns c) 3 turns and d) 4 turns MCR.

5. Summary

In the past, several fabrication techniques for tapered fibers/microfibers have been suggested. In this chapter, fabrication of microfiber based on flame brushing technique is reviewed. Flame brushing technique is commonly used for the fabrication of fiber couplers and tapered fibers. This technique enables fabrication of biconical tapered fibers which are important components for the manufacture of microfiber based devices. In order to achieve that, a fiber tapering rig was assembled. The heat source comes from an oxy-butane torch with a flame width of 1mm. Two stepper motors are incorporated in the rig to control the movement of the torch and translation stage. A biconical tapered fiber with a waist diameter as small as 700nm can be achieved with the rig. To achieve low loss tapered fibers, the shape of the taper should be fabricated according to adiabaticity criteria. In the laboratory, tapered fibers with linear and decaying-exponential profiles have been demonstrated. To provide protection to the tapered fibers, they are embedded in a low-index material or packaged into a perspex case. These protection measures can prolong the life span and stabilize the temporal performance of these tapered fibers and microfiber based devices.

Three microfiber based devices have been reviewed in this chapter namely MLR, MKR and MMZI. In the first section, the fabrication of MLR is introduced. MLR is assembled from a single mode microfiber by coiling it into a loop. A closed optical path within the loop is established when the two microfibers are put in close contact with each other and form an evanescent coupling. Then, the theoretical model of the self-touching MLR is presented and used for curve-fitting with the experimental data. The important characteristic parameters of the transmission spectra can be extracted from the best-fit curve in the experimental data. In the next section, MKR is presented. Similar to MLR, MKR shares the similar transmission characteristics and the same theoretical model for the MLR can be applied to MKR. However, MKR outperforms MLR in several aspects for instance MKR has more stable and stronger coupling due to its small spacing between the two coupling microfibers in the coupling region. In addition, the structure of MKR is more rigid and robust with the interfiber twisted coupling in the resonator. Nonetheless, MKRs suffer a setback in higher insertion loss if compared with MLRs because of the cut-coil-couple process in the fabrication. The evanescent coupling between output microfiber and collector microfiber contributes a large fraction in the total insertion loss. When MKR is embedded into a medium of different refractive index, the coupling in the MKR varies and it alters the resonance state of the resonator. By curving the experimental data for transmission spectra with the theoretical model, the coupling coefficient and round-trip attenuation factor of the MKR can be extracted from the best-fit curve. The results indicate that the state of resonance of an embedded MKR has been altered and it is closer to critical coupling condition.

Besides, microfiber based resonators exhibit an interesting polarization dependent characteristic in the investigation. The coupling coefficient in the resonator is dependent on input state of polarization and it can be manipulated by using a PBS and a PC. In relation with the coupling, the resonance extinction ratio (RER) of transmission spectrum can be varied by tuning the PC. In the experiment, the variation of RER between 5dB and 11dB was observed. In the past, several literatures have reviewed that the refractive index and dimension of silica microfiber have a strong dependence on temperature. The resonance extinction ratio and resonance wavelength can be expressed in functions of temperature. In order to gain insight into this characteristic, an investigation was conducted to study temperature response of an MKR. Both theoretical analysis and experimental result indicate that the spectral shift in the transmission of MKR is linearly proportional to the increment of temperature. This characteristic can be exploited in many applications particular in sector industrial sensors. By wrapping the microfiber knot on a conductor rod, MKR can perform as a current sensor. When an electric current is loaded through the conductor rod, the rod acts as a heating element and increases the temperature of the MKR. As a result, transmission spectrum of the MKR is shifted. This opto-electric configuration can be used a current sensor and an electric controlled optical filter. In the final section, the fabrication of MCR was demonstrated. An MCR of different number of turns was assembled and tested.

6. References

Bilodeau, F., K. O. Hill, S. Faucher, and D. C. Johnson, (1988), Low-loss highly overcoupled fused couplers: Fabrication and sensitivity to external pressure, *Lightwave Technology, Journal of* 6, 1476-1482.

Birks, T. A., and Y. W. Li, (1992), The shape of fiber tapers, *Lightwave Technology, Journal of* 10, 432-438.

Brambilla, G., F. Xu, and X. Feng, (2006), Fabrication of optical fibre nanowires and their optical and mechanical characterisation, *Electronics Letters* 42, 517-519.

Bricheno, T., and V. Baker, (1985), All-fibre polarisation splitter/combiner, *Electronics Letters* 21, 251-252.

Caspar, C., and E. J. Bachus, (1989), Fibre-optic micro-ring-resonator with 2 mm diameter, *Electronics Letters* 25, 1506-1508.

Chen, C.-L., and W. K. Burns, (1982), Polarization characteristics of single-mode fiber couplers, *Microwave Theory and Techniques, IEEE Transactions on* 30, 1577-1588.

Chen, Yi-Huai; Wu, Yu; Rao, Yun-Jiang; Deng, Qiang; Gong, Yuan, (2010), Hybrid mach-zehnder interferometer and knot resonator based on silica microfibers, *Optics Communications* 283, 4.

Ding, Lu, Cherif Belacel, Sara Ducci, Giuseppe Leo, and Ivan Favero, (2010), Ultralow loss single-mode silica tapers manufactured by a microheater, *Appl. Opt.* 49, 2441-2445.

Dong, H., G. Zhu, Q. Wang, H. Sun, N. K. Dutta, J. Jaques, and A. B. Piccirilli, (2005), Multiwavelength fiber ring laser source based on a delayed interferometer, *Photonics Technology Letters, IEEE* 17, 303-305.

Frazão, O., and et al., (2005), Chirped bragg grating fabricated in fused fibre taper for strain-temperature discrimination, *Measurement Science and Technology* 16, 984.

Graf, J. C., S. A. Teston, P. V. de Barba, J. Dallmann, J. A. S. Lima, H. J. Kalinowski, and A. S. Paterno, 2009, Fiber taper rig using a simplified heat source and the flame-brush technique, Microwave and Optoelectronics Conference (IMOC), 2009 SBMO/IEEE MTT-S International.

Guo, Xin, Yuhang Li, Xiaoshun Jiang, and Limin Tong, (2007), Demonstration of critical coupling in microfiber loops wrapped around a copper rod, *Applied Physics Letters* 91, 073512-3.

Guo, Xin, and Limin Tong, (2008), Supported microfiber loops for optical sensing, *Opt. Express* 16, 14429-14434.

Harun, S., K. Lim, A. Jasim, and H. Ahmad, (2010), Fabrication of tapered fiber based ring resonator, *Laser Physics* 20, 1629-1631.

Harun, S. W., K. S. Lim, A. A. Jasim, and H. Ahmad, (2010), Dual wavelength erbium-doped fiber laser using a tapered fiber, *Journal of Modern Optics* 57, 2111 - 2113.

Hou, Changlun, Yu Wu, Xu Zeng, Shuangshuang Zhao, Qiaofen Zhou, and Guoguang Yang, (2010), Novel high sensitivity accelerometer based on a microfiber loop resonator, *Optical Engineering* 49, 014402-6.

Hsu, Shih-Hsin, and Yang-Tung Huang, (2005), Design and analysis of mach?Zehnder interferometer sensors based on dual strip antiresonant reflecting optical waveguide structures, *Opt. Lett.* 30, 2897-2899.

Jiang, Xiaoshun, Limin Tong, Guillaume Vienne, Xin Guo, Albert Tsao, Qing Yang, and Deren Yang, (2006), Demonstration of optical microfiber knot resonators, *Applied Physics Letters* 88, 223501-223501-3.

Jung, Y, G. S. Murugan, G. Brambilla, and D. J. Richardson, (2010), Embedded optical microfiber coil resonator with enhanced high-q, *Photonics Technology Letters, IEEE* 22, 1638-1640.

Li, Yuhang, and Limin Tong, (2008), Mach-zehnder interferometers assembled with optical microfibers or nanofibers, *Opt. Lett.* 33, 303-305.

Lim, K. S., S. W. Harun, S. S. A. Damanhuri, A. A. Jasim, C. K. Tio, and H. Ahmad, (2011), Current sensor based on microfiber knot resonator, *Sensors and Actuators A: Physical* 167, 377–381.

Lim, K. S., S. W. Harun, A. A. Jasim, and H. Ahmad, (2011), Fabrication of microfiber loop resonator-based comb filter, *Microwave and Optical Technology Letters* 53, 1119-1121.

Liu, Zhihai, Chengkai Guo, Jun Yang, and Libo Yuan, (2006), Tapered fiber optical tweezers for microscopic particle trapping: Fabrication and application, *Opt. Express* 14, 12510-12516.

Love, J. D., W. M. Henry, W. J. Stewart, R. J. Black, S. Lacroix, and F. Gonthier, (1991), Tapered single-mode fibres and devices. I. Adiabaticity criteria, *Optoelectronics, IEE Proceedings J* 138, 343-354.

Mora, J., A. Diez, M. V. Andres, P. Y. Fonjallaz, and M. Popov, (2004), Tunable dispersion compensator based on a fiber bragg grating written in a tapered fiber, *Photonics Technology Letters, IEEE* 16, 2631-2633.

Ngo, N. Q., S. Y. Li, R. T. Zheng, S. C. Tjin, and P. Shum, (2003), Electrically tunable dispersion compensator with fixed center wavelength using fiber bragg grating, *Lightwave Technology, Journal of* 21, 1568-1575.

Orucevic, Fedja, Valérie Lefèvre-Seguin, and Jean Hare, (2007), Transmittance and near-field characterization of sub-wavelength tapered optical fibers, *Opt. Express* 15, 13624-13629.

Schwelb, O., (2004), Transmission, group delay, and dispersion in single-ring optical resonators and add/drop filters-a tutorial overview, *Lightwave Technology, Journal of* 22, 1380-1394.

Sumetsky, M., (2008), Basic elements for microfiber photonics: Micro/nanofibers and microfiber coil resonators, *J. Lightwave Technol.* 26, 21-27.

Sumetsky, M., Y. Dulashko, J. M. Fini, and A. Hale, (2005), Optical microfiber loop resonator, *Applied Physics Letters* 86, 161108-161108-3.

Sumetsky, M., Y. Dulashko, J. M. Fini, A. Hale, and D. J. DiGiovanni, (2006), The microfiber loop resonator: Theory, experiment, and application, *Lightwave Technology, Journal of* 24, 242-250.

Sumetsky, M., Y. Dulashko, and S. Ghalmi, (2010), Fabrication of miniature optical fiber and microfiber coils, *Optics and Lasers in Engineering* 48, 272-275.

Tong, Limin, Rafael R. Gattass, Jonathan B. Ashcom, Sailing He, Jingyi Lou, Mengyan Shen, Iva Maxwell, and Eric Mazur, (2003), Subwavelength-diameter silica wires for low-loss optical wave guiding, *Nature* 426, 816-819.

Vienne, G., Li Yuhang, and Tong Limin, (2007), Effect of host polymer on microfiber resonator, *Photonics Technology Letters, IEEE* 19, 1386-1388.

Vienne, Guillaume, Aurélien Coillet, Philippe Grelu, Mohammed El Amraoui, Jean-Charles Jules, Frédéric Smektala, and Limin Tong, (2009), Demonstration of a reef knot microfiber resonator, *Opt. Express* 17, 6224-6229.

Wu, Yu, Yun-Jiang Rao, Yi-huai Chen, and Yuan Gong, (2009), Miniature fiber-optic temperature sensors based on silica/polymer microfiber knot resonators, *Opt. Express* 17, 18142-18147.

Xu, Fei, and Gilberto Brambilla, (2007), Embedding optical microfiber coil resonators in teflon, *Opt. Lett.* 32, 2164-2166.

Xu, Fei, Gilberto Brambilla, and David J. Richardson, 2006, Adiabatic snom tips for optical tweezers.

Xu, Fei, Peter Horak, and Gilberto Brambilla, (2007), Optical microfiber coil resonator refractometric sensor, *Opt. Express* 15, 7888-7893.

Xu, Fei, Valerio Pruneri, Vittoria Finazzi, and Gilberto Brambilla, (2008), An embedded optical nanowire loop resonator refractometric sensor, *Opt. Express* 16, 1062-1067.

Yang, Szu-Wen, and Hung-Chun Chang, (1998), Numerical modeling of weakly fused fiber-optic polarization beamsplitters. I. Accurate calculation of coupling coefficients and form birefringence, *Lightwave Technology, Journal of* 16, 685-690.

Zeng, Xu, Yu Wu, Changlun Hou, Jian Bai, and Guoguang Yang, (2009), A temperature sensor based on optical microfiber knot resonator, *Optics Communications* 282, 3817-3819.

Zhang, J., P. Shum, X. P. Cheng, N. Q. Ngo, and S. Y. Li, (2003), Analysis of linearly tapered fiber bragg grating for dispersion slope compensation, *Photonics Technology Letters, IEEE* 15, 1389-1391.

Zhang, Rui, Jörn Teipel, Xinping Zhang, Dietmar Nau, and Harald Giessen, (2004), Group velocity dispersion of tapered fibers immersed in different liquids, *Opt. Express* 12, 1700-1707.

2

"Crystalline" Plastic Optical Fiber with Excellent Heat-Resistant Property

Atsuhiro Fujimori
Saitama University
Japan

1. Introduction

General "polymer crystals" essentially both crystalline and amorphous regions. It is well known that crystalline polymers construct hierarchical structures ranging from lamellae on the nanometer scale to spherulite on the mesoscopic scale.[1-3] The polymer crystals in these crystalline polymers are generally formed by the folding of the main chain. In many cases, since these folded parts and interspherulite chains form the amorphous region, crystalline polymers are essentially intermingled states of the crystalline and the amorphous regions. Therefore, crystalline polymers are not a suitable candidate for use in plastic optical fibers (POFs) and film-type optical waveguides (FOWs) because of the occurrence of light refraction at the crystalline/amorphous interface. Consequently, amorphous POFs lack heat resistance and dimensional stability.

However, if the construction of extremely homogeneous crystalline POFs is realized, "crystalline" POFs with excellent heat resistance and dimensional stability can be developed. The heat-resistant POFs will efficiently demonstrate their optical ability in a circuit exposed to a high temperature of more than 125 °C; so far there have been no products of heat-resistant POFs that can sustain temperatures higher than 125 °C. If the heat-resistant POFs are realized, light wiring in automobiles will also be achieved; the heat-resistant POFs will not only connect the AV equipment but also connect the control system around the engine. As a result, the overall body of an automobile will become lighter. This future technology is based mainly on "crystalline fluorinated polymers" having a high crystallinity. Generally, polytetrafluoroethylene (PTFE; $-(CF_2-CF_2)_n-$) and its copolymers easily form rigid helices in order to yield extended-chain crystals. It seems difficult for PTFE to form a lamellae structure because of its rigid molecular chain.[4-8] In addition, since tetrafluoroethylene copolymers obtained by the incorporation of several comonomers exhibit extremely fast crystallization rates,[9] their spherulites generally cannot be observed until they are sufficiently large. Therefore, PTFE exhibits a high degree of crystallinity of over 90%.[10-12]

Poly[tetrafluoroethylene-*co*-(perfluoroalkylvinylether)] (*abbrev*. EFA (alkyl = ethyl) or PFA (alkyl = propyl))[13] has a unique role in the plastics industry due to its inertness, heat resistance, and low coefficient of friction in a wide temperature range. Generally, fluorinated compounds and fluoropolymers have excellent chemical resistance, oil resistance, and oil- and water-shedding resistance.[14-17] They have been used as rubbers at high temperatures and in several lubricating fluorine manufactured products.

However, in the field of fundamental science, structural studies on fluorinated polymers have progressed slowly since the time these polymers were first reported by Bunn and Howells in 1954.[18] We could find very few reports on the systematic structural studies on PTFE or tetrafluoroethylene-based fluorinated copolymer because this compound is difficult to synthesize due to the emission of poisonous gases.[4, 6]

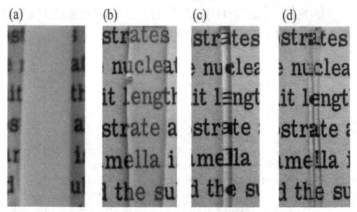

Fig. 1. Changes in transparency of several processed materials of "crystalline" fluorinated copolymers: (a) bulk EFA, (b) pressed processing sheet, (c) crystalline fiber with drawn ratio = 3, (d) crystalline fiber with drawn ratio = 5.

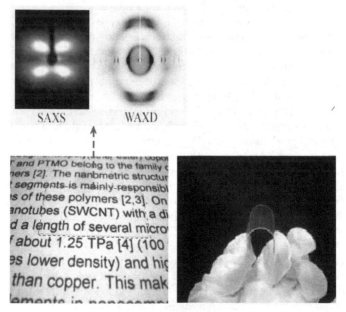

Fig. 2. Photograh of crystalline, transparent, and flexible film made by fluorinated copolymer, and their SAXS and WAXD patterns.

Further, although EFA is a crystalline polymer, processed EFA samples that have a high crystallinity are as transparent as amorphous flexible polymers such as polymethymethacrylate (PMMA)[19] and poly(ethylene terephthalate) (PET), as shown Fig. 1 and Fig. 2. This experimental fact is not well known. Probably, since the transparency of organic materials depends on the existence of differences in electron density between the crystalline and the amorphous regions, it is considered that a high crystallinity of EFA relates closely to the occurrence of transparency. Additionally, processed EFA tubes break into pieces just like glass tubes when an excessive bending force is applied upon them. It is obvious that the enhancement of these unique properties of the processed EFA POFs and FOWs is a result of the changes in the crystal structure and crystalline morphology of EFA fibers that take place during the drawing process. Further, fluorinated polymers do not absorb infrared light because of their stretching vibration and a lack of C-H bonds.[20, 21] Hence, a "crystalline" POF and FOW made by fluorinated polymers transports not only visible light but also infrared light.

In this chapter, the changes in the fine structure and lamella arrangement of the fibers formed by tetrafluoroethylene copolymers upon drawing are investigated by using wide-angle X-ray diffraction (WAXD) and small-angle X-ray scattering (SAXS) methods. We have found very few reports on the studies on the structural changes in fluorinated polymers upon drawing, whereas there are many reports of studies on hydrogenated polymers. Therefore, this study may also be valuable as fundamental research in the field of polymer physics. In addition, we have discussed the relationships between the origins in order to elucidate the occurrence of transparency and structural changes in molecular arrangements.

2. Experimental

2.1 Materials
2.1.1 Fluorinated copolymer
The fluorinated copolymers used in this study were provided by DuPont-Mitsui Fluorochemicals Co. Ltd. EFA is a random copolymer obtained from the copolymerization of tetrafluoroethylene $-(CF_2-CF_2)_n-$ and perfluoroethylvinylether $-(CF_2-CF(OCF_2CF_3))_n-$. The amount of comonomers of these materials was about 3 wt%. The molecular weight of the EFA processed to a crystalline fiber form was about 600,000. This molecular weight was examined by a computer simulation on the basis of the viscoelasticity of the fiber in a molten state because it is difficult to dissolve these polymers in an organic solvent.

2.1.2 Drawing of EFA POFs and FOWs
EFA POFs and FOWs were drawn uniaxially by using a hand-drawing apparatus in an air oven at 280 °C. The surface of the POFs and FOWs specimen was marked at intervals of 2 mm in order to measure the draw ratios. The drawing speed was fixed at 20 mm/min, and the fiber was annealed at 280 °C for 3 min before drawing. Using this method, we obtained fibers with excellent transparency (Figs. 1(c) and 1(d)).

2.2 Experimental methods
2.2.1 Small-angle X-ray scattering (SAXS)
The crystalline morphology of the drawn EFA copolymers was characterized with a SAXS instrument (M18XHF, MAC Science Co.) consisting of an 18-kW rotating-anode X-ray generator with a Cu target (wavelength, λ = 0.154 nm) operated at 50 kV and 300 mA.[22] This

instrument comprised a pyrographite monochromator, pinhole collimation system (ϕ ~0.3, 0.3, and 1.1 mm), vacuum chamber for the scattered beam path, and a two-dimensional imaging plate detector (DIP-220). The sample-to-detector distance was adjusted to 710 mm. The exposure time for each sample was 30 min. For the SAXS measurements, each sample (thickness: approximately 0.5 mm) was placed in a sample holder so that its position remained unchanged. The theoretical detection limit of the SAXS measurement in this study almost corresponded to the value of q = 0.128 nm^{-1} estimated by using the camera distance (from sample to the imaging plate) in the apparatus. However, the actual detection limit examined by counting the pixel numbers of enlarged SAXS patterns on the monitor of an analytical computer was q = 0.170 nm^{-1} (dashed line in the profile of Fig. 3). Hence, the observable maximum value of the long period between the centers of gravities of the lamellae in this study was 36.9 Å.

2.2.2 Wide-angle X-ray diffraction (WAXD)
In order to obtain the WAXD data for the drawn fibers, an R-axis diffractometer (Rigaku Co.) was operated at 45 kV and 200 mA to generate CuKα radiation (λ = 0.1542 nm). WAXD photographs of the samples were taken at room temperature by using a graphite monochromator and a 0.3-mm pinhole collimator. Diffraction data were recorded on a cylindrical imaging plate detector equipped with an interface to a computer system. The camera length was 127.4 mm, and the exposure time was 600 s.

2.2.3 Estimation of thermal properties and transparency
Thermal analyses were carried out by using a Seiko Instruments model DSC200 differential scanning calorimeter (DSC). The DSC measurements were performed at a standard scanning rate of 10.0 °C min^{-1}. A sample mass of about 5.00 mg was used for all the DSC measurements. As usual, the scanning of DSC measurements and the heating and cooling cycle were repeated twice in order to examine the difference between the peak position and transition enthalpy in the first and second heating. UV-vis spectra of EFA films were measured using a UV-vis spectrophotometer (V-650, JASCO).

3. Results and discussion

3.1 Changes in lamellae arrangement of transparent "crystalline" EFA POFs and FOWs
Figure 3 shows the SAXS pattern and normalized one-dimensional SAXS profiles, where q is the scattering vector (q = 4πsinθ/λ; θ = Bragg angle), of the undrawn transparent crystalline EFA POWs. A ring-shaped SAXS pattern was observed, which indicated the formation of an isotropic random lamella texture. In the case of PTFE, the SAXS pattern was obscure, and the corresponding profile exhibited extremely low intensity because this polymer almost formed an extended chain and not a lamellae structure.[23] On the contrary, it was found that the tetrafluoroethylene copolymer formed lamellae structures since the undrawn EFA used in this study exhibited isotopic SAXS patterns. The long period of the undrawn sample was estimated to be 27.0 nm. A high-crystallinity EFA sample formed relatively thicker lamellae than the general hydrogenated crystalline polymers.
On the basis of the results of the SAXS measurements of the undrawn EFA fiber, we suggested the following lamella model for tetrafluoroethylene copolymers. According to A. Keller's suggestion,[1] it was assumed that general crystalline polymers form a regular sharp

hold. However, the tetrafluoroethylene copolymers used in this study did not form an arrangement of these adjacent reentries because of the existence of a rigid molecular chain and a lack of flexibility. It seemed that the folded parts formed in the ether bond-rich region within the fluorinated main chain. However, so many perfluoroalkylvinylether units could not have contributed to the formation of the folded parts because the ratio of the absolute amounts of the comonomers was extremely low. Hence, we proposed a "switch-board type" lamellae model of these tetrafluoroethylene copolymers, shown in Fig. 4, according to P. J. Flory's suggestion.[23, 24] In this case, it was supposed that there existed a relatively large amorphous region because of the existence of the large long-period structure estimated by SAXS. From the qualitative estimation of the lamella thickness based on the crystallization degree obtained from the DSC measurements, the thickness of the crystalline regions of the EFA lamella form was estimated to vary within a range from 8 to 15 nm (as calculated by using the fusion enthalpy of as-polymerized PTFE, ΔH_{endo} (58.4 J g^{-1}), as the standard fusion enthalpy of EFA, $\Delta H_{endo, 0}$).[23] The existence of the thick amorphous layer (over 10 nm) also supports the validity of our proposed switch-board type lamella model.

Figure 5 shows the SAXS patterns and corresponding lamella arrangement models for DR1 (draw ratio = 1.0, undrawn), DR3, and DR5 transparent crystalline POFs of EFA. A ring-shaped SAXS pattern was observed for the undrawn DR1 sample (Fig. 5 (a)), while two- or four-point patterns were observed for the DR3 (Fig. 5 (b)) or DR5 (Fig. 5 (c)) fiber samples. The former indicated a random lamellar texture (Fig. 5 (a')), and the latter indicated some lamella structures oriented with respect to the draw direction.

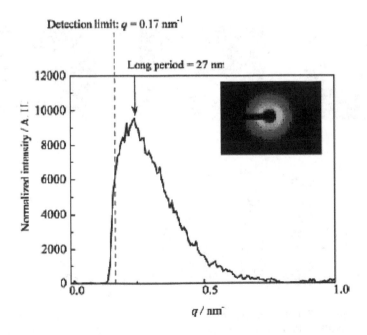

Fig. 3. SAXS pattern and profile of undrawn EFA 'crystalline' POF.

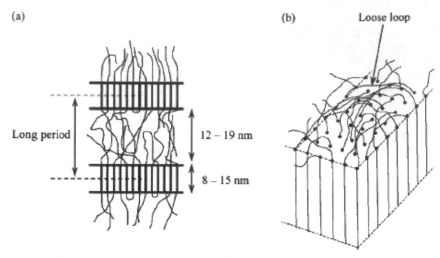

Fig. 4. Schematic illustrations of "switchboard-type" lamella models of fluorinated copolymers like an EFA (a) along the *c*-axis, and (b) in an *a-b* plane.

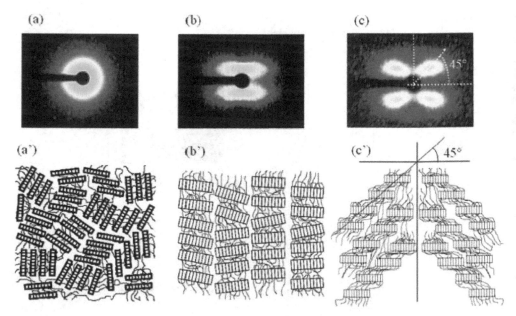

Fig. 5. Changes in SAXS patterns and corresponding lamella arrangement models of EFA transparent 'crystalline' POF with drawing; (a), (a') undrawn, (b), (b') 3 times, and (c), (c') 5 times drawing.

The appearance of the two-point SAXS patterns implied the formation of an arrangement of lamellae parallel to the draw direction (Fig. 5 (b')). As the fiber was drawn further, the interlamella and/or intralamella slips probably occurred, forming the small kink bands in the lamellae. The slip angle of the interlamellae was 45° as calculated by using the position of the strongest spot in the SAXS picture. In accordance with the changes in lamellae, the grain boundaries or amorphous parts between two neighboring lamellae were also distributed regularly towards the draw direction, and they thus resulted in a periodic change in density in the direction normal to them, which accounted for the four-point diffraction pattern. That is, with an increase in the elongation of the EFA sample, a particular kind of layer structure, an alternately tilted lamella arrangement known as the herringbone, was formed inside the fibers (Fig. 5 (c')). Similar results were obtained in the case of drawn polyethylene (PE) fibers previously.[25] The long periods or interplanar spacings were calculated to be 33.9 and 35.3 nm for DR3 and DR5, respectively. These values were larger than the interplanar spacing of the undrawn sample (27.0 nm). This feature of the long periods corresponded well with that of PE, polypropylene (PP), and polyester.[25-30] From the viewpoint of enhancing transparency by using the drawing process, EFA fibers exhibited the elongation of the amorphous region with an increase in density in this region and indicated a resultant increase in the long period upon drawing.

Figure 6 shows the change in SAXS patterns upon drawing. SAXS patterns remained essentially unchanged even upon carrying out the drawing process for five times. However, from the results of the examination of light conductivity in db/km units for EFA fibers using infrared light (at λ = 850 nm), most superior abilities were confirmed in the DR5 fibers, and their transmission ability was observed to decrease gradually upon drawing for over six times. Moreover, the drawn EFA fiber broke when the elongation equaled almost nine times the original value. Just before breaking, the color of the drawn EFA fiber became white because of the appearance of many microvoids and/or defects and the light dispersion caused by these voids and/or defects. In order to estimate the changes in lamella thickness and differences in electron density upon drawing, plots of the draw ratio *vs.* long periods and normalized intensity of SAXS profiles are shown in Fig. 7. The values of the long period saturated at about DR3, and the normalized intensity was almost constant from DR4 to DR8. That is, the increase in the lamella thickness containing an amorphous region stopped at DR3 (about 35 nm). After that, although the density of the amorphous region increased gradually upon drawing, a partial appearance of the voids might have occurred simultaneously. As a result, the difference in the overall density between the crystalline and the amorphous regions in the EFA fiber remained unchanged for a draw ratio of more than 4.

3.2 WAXD study on crystal structure of tetrafluoroethylene-based polymers

A typical example of the WAXD patterns for the drawn EFA fibers (DR8) is shown in Fig. 8(a). Almost all spots existed on the equator line. Therefore, we have mainly discussed the WAXD profiles integrated along the equatorial direction in this section. Figure 8(b) shows a comparison of the WAXD profiles of the unoriented PTFE and the EFA samples. The lack of an amorphous curve around 2θ = 15° was a peculiarity of the PTFE extended-chain crystal. A halo curve of the EFA appeared due to the existence of an amorphous region in the interlamella parts. However, the crystalline peak positions in both profiles were almost the same since the structure and main-chain arrangement in the crystalline region of EFA

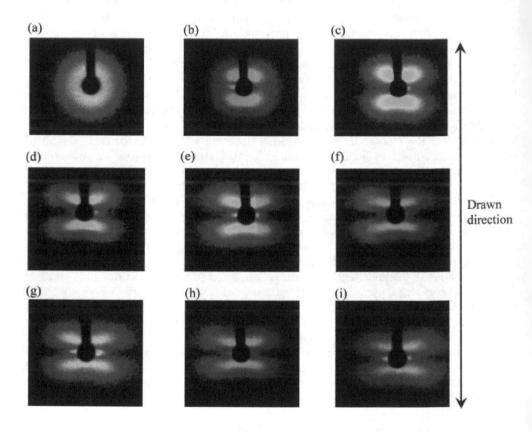

Fig. 6. Changes in SAXS patterns of EFA "crystalline" POFs with drawing at a ratio of (a) 1.0, (b) 1.5, (c) 2.0, (d) 3.0, (e) 4.0, (f) 5.0, (g) 6.0, (h) 7.0, and (i) 8.0.

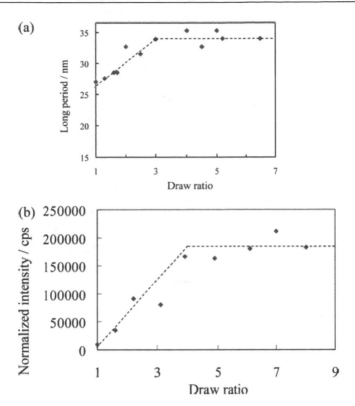

Fig. 7. Plots of draw ratio vs (a) long periods and (b) normalized intensity estimated by SAXS measurements of EFA "crystalline" POFs.

comprised repeating tetrafluoroethylene parts. That is, there was no difference between the structure of the crystalline region of PTFE and that of EFA.

Furthermore, most inner WAXD spots of an EFA fiber (Fig. 8(a); the shadow next to the beam stopper) existed clearly when $2\theta = 9°$. These WAXD results included a very important result with regard to the fluorinated polymer crystal. The peak at around $2\theta = 18.0°$ in the WAXD profiles of tetrafluoroethylene and its copolymers was assigned to the (100) reflection in the quasi-hexagonal system according to the literature documented about 50 years ago.[18, 31-33] Moreover, we could not find any reports related to the inner peak around $2\theta = 9°$. However, in the present WAXD profiles, small peaks at around $2\theta = 9°$ were confirmed and reproduced well by the high-power measurement using an X-ray diffractometer with an imaging plate as the detector. Further, in the WAXD profile of the oriented rod-shaped material processed by isostatic pressing of PTFE, this peak was clearly enhanced (Fig. 8(c)). In addition, Fig. 9 shows the changes in this peak in the WAXD profiles of the transparent crystalline EFA fiber upon drawing and the well-reproduced appearance of this peak in any type of fluorinated copolymers. From the result of Fig. 9(a), it was found that the intensity of this peak around $2\theta = 9.0°$ increased gradually with an increase in the draw ratio. Figure 9(b) shows the WAXD profiles of several fluorinated copolymers such as PTFE, poly[tetrafluoroethylene-co-(hexafluoropropylene)] (FEP), PFA, PFA containing PTFE

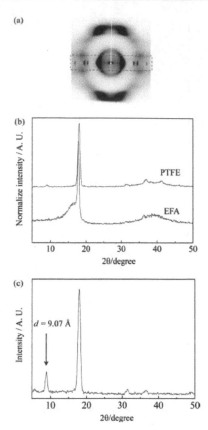

Fig. 8. (a) WAXD patterns of "crystalline" POFs of EFA at draw ratio = 8. (b) Comparison of WAXD profiles of EFA to PTFE. (c) WAXD profile of PTFE orientated rod formed by isostatic extrusion.

particles as nucleators, low molecular weight EFA (250,000), middle molecular weight EFA (300,000), and high molecular weight EFA (600,000) containing PTFE particles. All WAXD profiles of fluoropolymers used in this study contained this small peak at almost the same position. That is, this small peak around $2\theta = 9.0°$ reflected that the genuine crystal structure of fluorinated polymers was always confirmed in the WAXD profiles of tetrafluoroetthylene-based polymers. Furthermore, the intensity of this peak increased upon the formation of an orientated structure due to uniaxial drawing. However, no previous reports that confirm the presence of these small peaks exist, except for the paper we published recent year.[22] It appears that the existence of this diffraction peak has been overlooked for about 50 years. In our previous report, we speculated that the peak at about $2\theta = 9°$ might correspond to the genuine (100) reflection.[23] In the present report, we clearly assert an interpretation of this peak and the crystal structure and partially modify our previous interpretation. In our previous work,[23] we suggested that the previously reported lattice constant needed to be modified and the lengths of the a- and b-axes be doubled. In addition, the reflection at around $2\theta = 18.0°$ would be attributed to the (200) peak. If this did

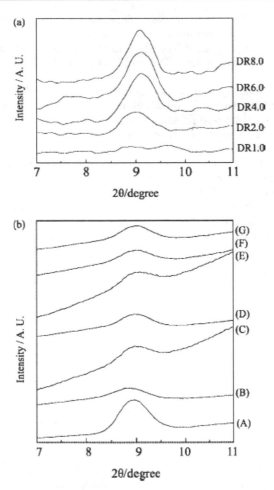

Fig. 9. WAXD patterns of (a) drawn EFA POFs, and (b) several fluorinated polymers in bulk: (A) PTFE, (B) FEP, (C) PFA, (D) PFA containing PTFE particle as nucleator, (E) low molecular weight EFA, (F) high molecular weight EFA, (G) EFA containing PTFE particle as nucleator.

not occur, the reflective indexes of the small peaks at about $2\theta = 9°$ could not be determined. Figures 10(a) and 10(b) show the reciprocal lattice of PTFE and other perfluorinated copolymers observed along the c-axis under the suggestion that the parts forming the crystal region had the same structure for tetrafluoroethylene and tetrafluoroethylene copolymers. The proposed lattice constant of PTFE[23] corresponded to $a = b = 11.08$ Å, $c = 16.8$ Å, $\alpha = 90°$, $\beta = 90°$, and $\gamma = 119.3°$ (Fig. 8(b), quasi-hexagonal system) and improved upon the reports by Bunn, et al., Starkweather Jr., et al., Clark, et al.,[18, 31-33] and other investigation groups (Fig. 10(a), $a = b = 5.54$ Å, $c = 16.8$ Å, $\alpha = 90°$, $\beta = 90°$, and $\gamma = 119.3°$ (quasi-hexagonal system)). However, the reciprocal lattice of Fig. 9(b) described a base-centered hexagonal lattice, whereas a base-centered lattice cannot exist in a group of hexagonal lattices. In addition, the reason for the appearance of a (100) reflection (peak at $2\theta = 9°$) weaker than a

(200) one (peak at 2θ = 18°) was not clear. Therefore, we reproposed the necessity of modifying the lattice constant of tetrafluoroethylene and its copolymers in the present work. We reconsidered the packing mode of fluorinated chains from a hexagonal to an orthorhombic system, as shown in Figs. 10(c) and 10(d). In the reciprocal lattice in Fig. 10(c), all WAXD reflection peaks confirmed in this study existed at a point of intersection in reciprocal lattice and all reflective indexes were decided. In this case, the peaks at 2θ = 9° and 18° corresponded to the (100) and (110) reflection peaks, respectively. The lattice constants of this packing system were estimated to be a = 9.58 Å, b = 5.54 Å, and c = 1.69 Å ($α$ = $β$ = $γ$ = 90°). The hexagonal lattice essentially had the structural analogy of an orthorhombic one. In addition, the appearance of peaks at 2θ = 9° and 18° was based on a different plane. Hence, the relation between intensities was not contradictory to an indexing rule. The three-dimensional packing model of the fluorocarbon chain in the crystalline region is shown in Fig. 10(d). The validity of our proposed orthorhombic system of the crystalline fluorinated polymer was also supported by the estimation in a reciprocal lattice along the meridional direction. Figure 11 shows the possibility for applying an orthorhombic lattice to an index WAXD reflection spot along the meridional direction of the drawn EFA fiber at DR5. As mentioned above, we considered the EFA chains as an orthorhombic packing in the crystal region, and the highest diffraction peak in the profile was interpreted as a (110) reflection in this lattice in the following discussion.

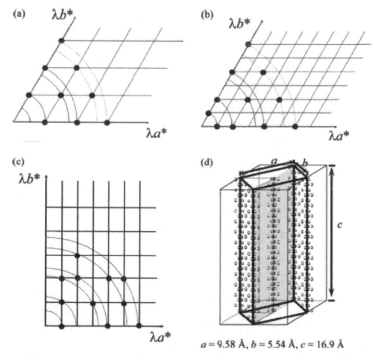

a = 9.58 Å, b = 5.54 Å, c = 16.9 Å

Fig. 10. Reciprocal lattices of crystalline region for several fluorinated polymers (PTFE, EFA, and so on) represented by WAXD data: (a) previously reported quasi-hexagonal lattice, (b) a quasi-hexagonal lattice twice elongated a- and b-axis, (c) our proposed orthorhombic lattice, and (d) packing model of fluorinated chains in orthorhombic lattice.

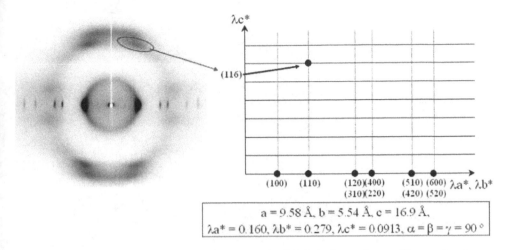

Fig. 11. WAXD patterns and reciprocal lattice in the λb^*-λc^* plane of crystalline region for EFA transparent fiber at DR5.

3.3 Fine structure estimation of transparent crystalline EFA POFs and FOWs upon drawing at subnanometer scales by WAXD

Figure 12 shows the WAXD patterns of the transparent EFA fiber at several drawing ratios. We can clearly see the gradual enhancement of the WAXD spots along the equator line upon drawing. From the viewpoint of one-dimensional profiles scanned along the equatorial direction, the peak intensity of (110), (120), (220), and (420) reflections in the orthorhombic lattice increased gradually with an increase in draw ratio (Fig. 13(a)). The intensities of (110) peaks normalized by sample size and thickness almost saturated at DR5, as observed from the plot of Fig. 13(b) whereas the sizes of crystallite in the fiber estimated by Schereer's formula[34] are almost constant value all over the draw ratio. That is, it was considered that the increase in the crystallinity of the EFA fiber at the subnanometer scale actually reached a constant value.

In order to evaluate the degree of orientation for the c-axis of the EFA crystallites along the draw direction, we calculated the orientation function (f) proposed by Hermans and co-workers[35] using the azimuthal WAXD profiles. The function f was defined as

$$f_\varphi = \frac{1}{2}(3 < \cos^2 \varphi > -1)., \quad 0 < f_\phi < 1,$$

where φ is the angle between the c-axis and the draw direction, and $<\cos^2\varphi>$ is obtained from the (110) and (120) azimuthal profiles by using Wilchinsky's procedure[36] (Fig. 14(a)). Figure 14(b) shows the change in the orientation function of the EFA crystallites ($f\varphi$) as a function of the draw ratio, where f_φ increased with the draw ratio up to DR = 2.5, after which it reached a saturation value of around 0.8. These findings suggested that the orientation of an EFA crystallite in the fiber was complete at a draw ratio of 2.5. This value was almost the same as the draw ratio of the saturation value of a long period estimated by SAXS. That is, the orientation of the crystallite and the elongation of lamella reached

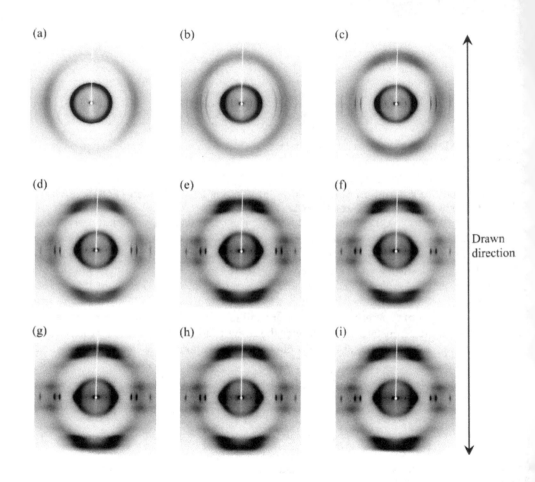

Fig. 12. WAXD patterns of EFA plastic optical fiber at several drawing ratio at room temparture: (a) undrawn, (b) DR) 1.5, (c) DR) 2.0, (d) DR) 3.0, (e) DR) 4.0, (f) DR) 5.0, (g) DR) 6.0, (h) DR) 7.0, (i) DR) 8.0.

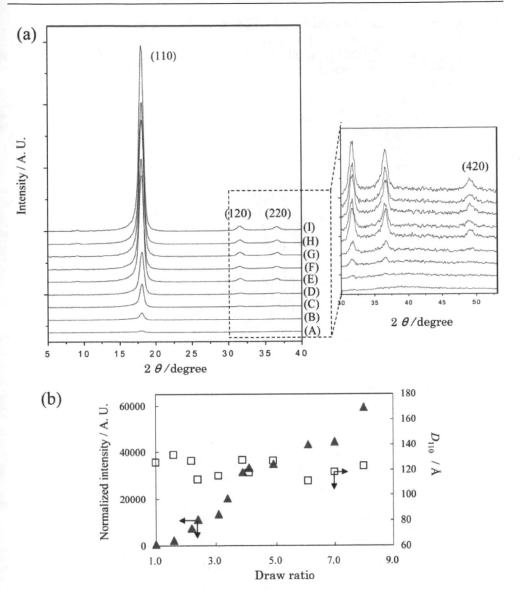

Fig. 13. (a) WAXD profiles of EFA plastic optical fiber with drawing at room temparture: (A) undrawn, (B) DR1.5, (C) DR2.0, (D) DR3.0, (E) DR4.0, (F) DR5.0, (G) DR6.0, (H) DR7.0, (I) DR8.0. (b) Changes in normalized WAXD intensity and crystallite sizes with drawing estimated by Scherrer's formula.

constant values almost simultaneously. Then, the quasi-crystallization process by drawing progressed up to DR5, which was the saturation value of the normalized intensity estimated on the basis of the WAXD patterns. Judging from the draw ratio of the saturation of the SAXS intensity, the increase in the electron density of the amorphous region and the partial appearance of voids might be a simultaneous occurrence upon further drawing. The sample of the crystalline EFA fiber at DR5 was the most transparent and exhibited the highest conductivity of infrared light among all the drawn fibers used in this study. In conclusion, the functionality of light transmittance was closely related to the solid-state structure of the crystalline EFA fiber.

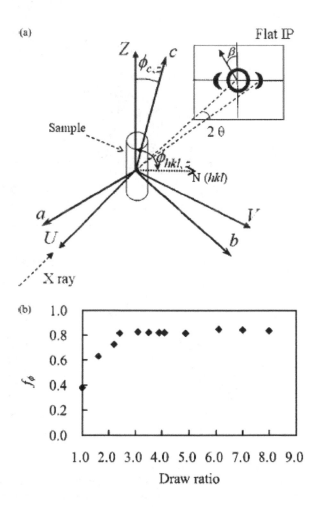

Fig. 14. (a) Schematic representation of Wilchinsky method to estimate orientation coefficient of crystallite. (b) Plot of drawn ratio vs orientation coefficient of crystallite in EFA POF.

Figure 15(a) shows results of DSC measurements of EFA "crystalline" fibers at several drawn ratios in order to estimate crystallization degree and lamella thickness. Areas of melting peaks on thermograms related to fusion enthalpy are gradually increased with drawn ratios. Crystallization degree as calculated by using the fusion enthalpy of as-polymerized PTFE, ΔH_{endo} (58.4 J g-1), as the standard fusion enthalpy of EFA, $\Delta H_{endo,0}$,[25] are plotted to drawn ratios of EFA fibers (Figure 15(b)). The linearity of changes in crystallinity of drawn fiber wellcorresponds to dependency of WAXD (110) intensity on drawing (Figure 12(b)). Further, from the qualitative estimation of the lamella thickness based on the crystallization degree, the thickness of the crystalline regions of the EFA lamella form was estimated to vary within a range from 6 to 16 nm (Figure 15(c)). In the case of DR5 fiber with most superior transmission ability of infrared light, almost 50% crystallinity and 11 nm lamella thickness are estimated. Therefore, it seems that the enhancement of transmission ability is not caused by increases of crystallinity, but reducing of differences in density between crystal and amorphous region. Probably, a high light transmission rate is not brought about by formation of extreme homogeneous crystalline fiber, but by formation of like a "fringed micelle-type" lamella arrangement which has an indistinct lamella-interface based on the enhancement of density for amorphous parts by drawing. In the case over six times drawing, since transition from amorphous part to crystalline part occurrs in EFA fiber, the density reduction of amorphous region and increases of differences in density between crystal and amorphous parts have developed. As a result, it seems that the transmission ability of infrared light decreases over six times drawing to EFA fibers.

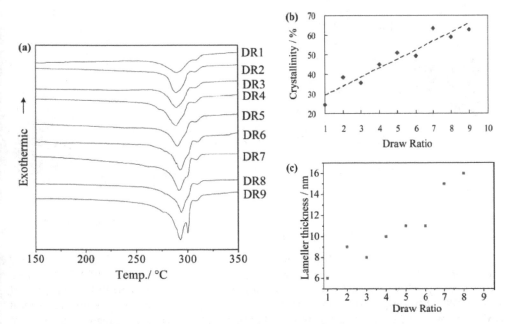

Fig. 15. (a) DSC thermograms of drawn EFA POFs at several ratios (scanning rate, 10 °C min-1). (b) Plot of drawn ratio vs crystallinity of drawn EFA fibers at several ratios. (c) Plot of lamellar thickness vs crystallinity of drawn EFA fibers at several ratios.

Figure 16 shows the schematic illustrations of the hierarchical structures ranging from the lamellae on the nanometer scale to the crystal structure on the subnanometer scale of a transparent EFA fiber.[37] We suggested that the crystal structure of the crystalline fluorinated polymers such as PTFE, EFA, PFA, and so on, form the orthorhombic system. The crystalline fiber of EFA had a herringbone arrangement in lamella when it was drawn over five times. Upon further drawing, the density in the amorphous region increased gradually. However, the overall differences in electron density between the crystalline and the amorphous regions were almost invariable. Probably, the progression of further transparency and the ability of light conductivity were brought about by a reduction in the difference in density. As an ideal type of extremely transparent crystalline fiber, the formation of a fringed micelle-type lamella arrangement may be desirable because of the low differences in densities inside the fibers.

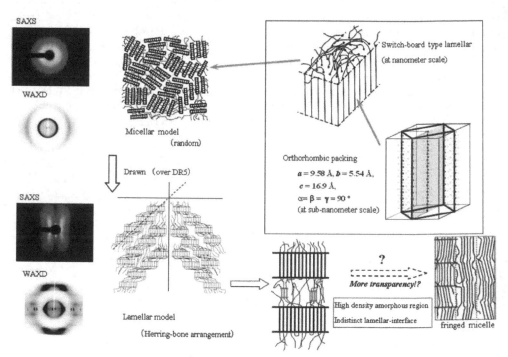

Fig. 16. Schematic illustrations of hierarchical structures from lamellae on the nanometer scale to crystal structure on the subnanometer scale of EFA POF.

Finally, in order to estimate three-dimensional structural formation, SAXS and WAXD measurements from the several incident direction of piled up crystalline EFA FOWs were carried out by using annealed DR=3 sample. Figure 17 shows SAXS and WAXD patterns of EFA FOWs at through, side, and edge direction. At the side-direction, obscure four-point SAXS pattern with void scattering and WAXD fiber pattern were confirmed. In the case of edge-direction, SAXS patterns show only void scattering, and WAXD indicate isotropic Debye ring. From the results of these measurements, schematic illustration of three-dimensional lamella arrangement was shown in Fig. 18. In this case, according to our previous work,[23, 38, 39] "switch-board" type lamella was adopted as structural units. From the view of through and side direction, two-dimensional stacked lamella arrangement forms the "herring-bone" arrangement. However, randomly isotropic structure is observed from edge direction. That is to say, lamella in the drawn EFA films formed uniaxially cylindrical symmetric arrangement. In the case of using this type EFA film as FOWs, it supposes that anisotropy of light conductivity direction occur. Along the through and side direction, visible and infrared light will be efficiently conducted, while edge direction will impede the transmission of lights. Figure 19 shows quantitative data of the transparency of the undrawn EFA film and drawn films by using UV-isible spectrometer. Because a "crystalline" FOWs made by fluorinated polymers efficiently transports infrared light, the λ= 850 nm of wavelength is adopted in this estimation. The film thickness is normalized by 500 µm. The transparency of infrared light in this film linearly increases with drawing ratio in both cases of films with drawing at 200 °C and fixed annealing at 280 °C after drawing. However, transparency of films treated by fixed annealing method is always inferior to that of films drawn at 200 °C only. This result is based on the difference of electron density between crystal and amorphous region. Probably, fixed annealing contributes acceleration of transition from a part of amorphous region to the crystal region. Crystallization of amorphous parts brings about formation of lower density amorphous region. As a result, difference of density between crystal and amorphous region become large and transparency of films decreases.

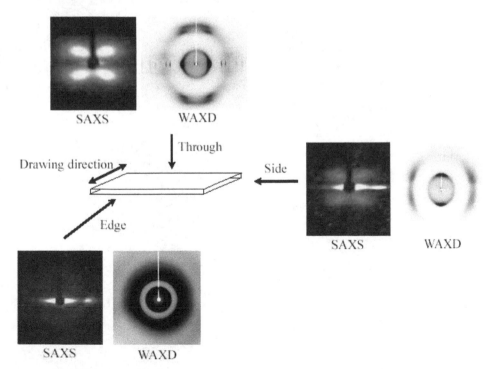

Fig. 17. SAXS and WAXD patterns of drawn EFA FOWs (fixed annealing at 280 °C after drawing at 200 °C) with through, side, and edge direction.

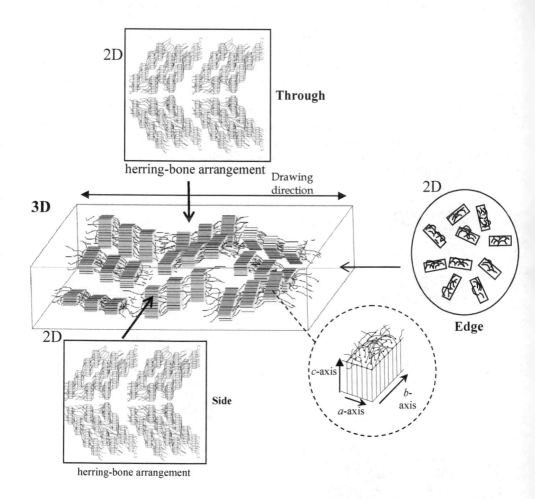

Fig. 18. Illustration of stacked lamellar in drawn EFA FOWs (fixed annealing at 280 °C after drawing at 200 °C).

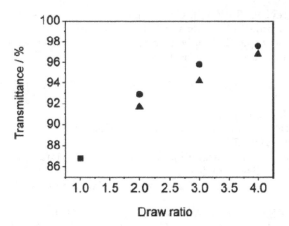

Fig. 19. Plots of drawing ratio *versus* transparency of infrared right (λ = 850 nm) : ■, undrawn ; ●, drawn at 200 °C; ▲, fixed annealing at 280 °C after drawing at 200 °C.

4. Conclusion

The changes in fine structure upon drawing transparent crystalline EFA fibers and films were investigated by WAXD and SAXS measurements. EFA was crystallized as a lamella crystal in the POFs and FOWs although the polytetrafluoroethylene homopolymer itself usually forms extended-chain crystals. EFA exhibited thicker lamellae (thickness: at least 27 nm) as observed by the SAXS measurement. In this type of crystalline fluorinated copolymers, we considered the formation of a switchboard-type lamellae model according to Flory's suggestion. With an increase in the drawing of the fibers and films, four-point SAXS diagrams developed in the photograph of EFA transparent fibers, which implied that a particular type of layer structure, the alternately tilted lamella arrangement known as the herringbone, was formed. Furthermore, it was found that the previously proposed packing mode of general fluorinated polymers was required to be reconsidered from quasi-hexagonal to orthorhombic in a reciprocal lattice in order to assign all the reflective indexes obtained by using high-resolution WAXD measurements. Furthermore, the orientation of the crystallite and the elongation of lamella of EFA were completed simultaneously in the drawn fibers. The quasi-crystallization process progressed upon further drawing up to five times. After that, an increase in the density of the amorphous region and a partial appearance of voids probably occurred simultaneously. The crystalline EFA fiber at DR5 exhibited excellent transparency and infrared light conductivity. The light transmission property was related closely to the lamella arrangement, crystal structure, and difference in the crystalline/amorphous density of crystalline EFA optical fibers and optical waveguide.

5. References

[1] Keller, A., *Phil. Mag.*, 1957, 2, 1171.
[2] Till, P. H., *J. Polym. Sci.*, 1957, 24, 301.
[3] Fischer, E. W., *Z. Naturforsch*, 1957, 12(a), 753.
[4] Burdon, J.; Tatlow J. C.; "*Advances in Fluorine Chemistry*," Vol.1 (eds. M. Stacey, J. C. Tatlow, A. G. Sharp, Academic Press, New York, 1960), pp.129–165.

[5] Patrick, C. R.; Stacey, M.; Tatlow, J. C.; Sharpe, A. G., "*Advances in Fluorine Chemistry,*" Vol. 2, Butterworths Publications Ltd., London, 1961, pp.1–34.

[6] "*Fluoropolymers 2*" in Topics in Applied Chemistry, eds. by Hougham, G., et al. (Kluwer Acad. / Plenum Pub., New York, 1999).

[7] Symons, N. K. J., *J. Polym. Sci., A*, 1963, *1*, 2843.

[8] Rahl, R. J.; Evanco, M. A.; Frendericks, R. J.; Reimschuessel, A. C., *J. Polym. Sci. A-2*, 1972, *1*, 1337.

[9] Ozawa, T., *Bull. Chem. Soc. Jpn.*, 1984, *57*, 952.

[10] Marega, C.; Marigo, A.; Causin, V.; Kapeliouchko, V.; Nicoló, E. D.; Sanguineti, A., *Macromolecules*, 2004, *37*, 5630.

[11] Marega, C.; Marigo, A.; Garbuglio, C.; Fichera, A.; Martorana, A.; Zannetti, R., *Macromol. Chem.*, 1989, *190*, 1425.

[12] Marega, C.; Marigo, A.; Cingano, G.; Zannetti, R.; Paganetto, G., *Polymer*, 1996, *37*(25), 5549.

[13] Lee, J. C.; Namura, S.; Kondo, S.: Abe, A., *Polymer*, 2001, *42*, 8631.

[14] Overney, R. M.; Meyer, E.; Frommer, J.; Brodbeck, D.; Luthi, R.; Howald, L.; Güntherodt, H. J.; Fujihira, M.; Takano, H.; Gotoh, Y., *Nature*, 1992, *359* 133.

[15] Overney, R. M.; Meyer, E.; Frommer, J.; Güntherodt, H. J., *Langmuir*, 1994, *10*, 1281.

[16] Fujimori, A.; Shibasaki, Y.; Araki, T.; Nakahara, H.,*Maclomol. Chem. Phys.*, 2004, *205*, 843.

[17] Fujimori, A.; Araki, T.; Nakahara, H., *Chem. Lett.*, 2000, 898.

[18] Burn, C. W.; Howells, E. R., *Nature*, 1954, *18*, 549.

[19] Koike, Y., Polymer, 1991, *32*, 1737.

[20] Koike, Y.; Naritomi, M.; Japan Patent 3719733, US Patent5783636, EU Patent 0710855, KR Patent, 375581, CN Patent ZL 951903152, TW Patent 090942, 1994.

[21] Ishigure, T.; Kano, M.; Koike, Y., *J. Lightw. Technol.*, 2000, *18*, 178.

[22] Nam, P. H.; Ninomiya, N.; Fujimori, A.; Masuko, T.; *Polym. Eng. Sci.*, 2006, *46*(6), 703.

[23] Fujimori, A.; Hasegawa, M.; Masuko, T., *Polym. Int.*, 2007, *56*, 1281

[24] Flory, P. J., *J. Am. Chem. Soc.*, 1962, *84*, 2857.

[25] Tanaka, K.; Seto, T.; Hara, T.; Tajima, Y., *Rep. Prog. Polym. Phys. Jpn.*, 1964, *7*, 63.

[26] Kaji, K.; Mochizuki, T.; Akiyama, A.; Hosemann, R., *J. Mater. Sci.*, 1978, *13*, 972.

[27] Samuels, R. J., *J. Macromol. Sci.*, 1970, *701*, 241.

[28] Butler, M. F.; Donald, A.N., *J. Appl. Polym. Sci.*, 1998, *67*, 321.

[29] Stribeck, N.; Sapoundjieva, D.; Denchev., Z.; Apostolov, A. A.; Zachmann, H. G.; Stamm, M.; Fakirov. S.; *Macromolecules*, 1997, *30*, 1339.

[30] Hernández, J. J.; Gracía-Gutiérrez, M. C.; Nogals, A.; Rueda, D. R.; Sanz. A.; Sics, I; Hsiao, B. S.; Roslaniec, Z.; Broza, G.; Ezquerra, T. A., Polymer, 2007, *48*, 3286.

[31] Sperati, C. A., Starkweather, H. W. Jr., *Adv. Polym. Sci.*, 1961, *2*, 465.

[32] Burn, C. W., Cobbold, A. J., Palmer, R. P., *J. Polym. Sci.*, 1958, *19*, 365.

[33] (a) Clark, E. S. Muus, L. T., *Z. Krist.*, 1962, *117*, 119, (b) Clark, E. S., Muus, L. T., *Z. Krist.*, 1962, *117*, 108.

[34] Klug, H. P.; Alexander, L E., *X-ray Diffraction Procedures*, John Wiley and Sons, New York, 1974.

[35] (a) Hermans P. H.; Platzek, P., *Kolloid Z.*, 1939, *88*, 68, (b) Hermans, J. J.; Hermans, P. H.; Vermaas, D.; Weidinger, A., *Rec. Trav. Chim, Pays-Bas*, 1946, *65*, 427.

[36] Wilchinsky, Z. W., *J. Appl. Phys.*, 1959, *30*, 792.

[37] Hayasaka, Y.; Fujimori, A., *Trans. Mater. Res. Soc. Jpn.*, 2008, *33*, 83-86.

[38] Fujimori, A.; Hayasaka, Y,, *Macromolecules*, 2008, *41*, 7606.

[39] Fujimori, A.; Numakura, K.; Hayasaka, Y, *Polym. Eng. Sci.*, 2010, *50*, 1295.

3

Design and Characterization of Single-Mode Microstructured Fibers with Improved Bend Performance

Vladimir Demidov, Konstantin Dukel'skii and Victor Shevandin
S.I. Vavilov Federal Optical Institute, St. Petersburg
Russia

1. Introduction

Over the last few years, clear progress has been made in research and development of single-mode optical fibers with a large core (when core diameter exceeds 10 μm). Such advances were stimulated essentially by growing requirements for means of high power laser radiation transmission. The urgent problem of laser beam delivery lies in the necessity of the primary Gaussian power distribution of light inherent to many laser sources to be maintained without both temporal and spatial distortions. So optical fibers that support only a single transverse mode prove to be the most appropriate technique for efficient light transfer in production areas of complex or compact architecture. But there are still a number of limitations to cope with. For instance, as the power density of generated laser beams increases, the fiber core has to be expanded adequately in order to minimize the impact of undesirable nonlinear effects such as Raman scattering, Brillouin scattering and self-phase modulation. Moreover, fiber material will exhibit irreversible breakdown if the power level equals or exceeds the critical damage threshold.

Conventional single-mode fibers with step-index or graded-index refractive index profile can be acceptably adapted for the realization of large cores. However, the core dimensions enlargement permanently results in the reduction of the refractive index difference between the core and the cladding (Δn). This, in turn, affects adversely the numerical aperture of the fiber (NA), which then has to be reduced twice from its standard values of larger than 0.1 to achieve core diameters of approximately 15 μm at a wavelength around 1 μm (Tunnermann et al., 2005). Such NA lowering weakens considerably the fiber waveguiding so the optical fiber becomes very sensitive to various perturbations, especially to bending effects. Further decrease of NA will require keeping the uniformity of the core refractive index in the vicinity of $10^{-4} - 10^{-5}$. It is technologically unattainable when using chemical vapor-phase deposition methods for the fiber preform fabrication.

An alternative flexible approach to solve this challenge is based on exploiting unique wave guiding properties of microstructured optical fibers (MOFs), also known as photonic crystal fibers or holey fibers. MOF design can relatively easily provide extended cores and hence large effective mode areas that nowadays reach values of even thousands of μm^2. This phenomenon perfectly coordinates with the ability to manage accurately the effective Δn value at a level of as low as 0.0001 or less. Furthermore, MOFs, as opposed to single-mode fibers of a conventional design, can assure robust fundamental mode propagation over a broad wavelength range within the transparency window of silica. The only restriction has to be taken into account while manufacturing microstructures with a large core relates to reasonable control over spectral position of bend-loss edge crucial for the fiber application potential.

Typical silica-based MOF structure is defined by a certain number of air holes arranged in a regular triangular lattice running along the entire length of the fiber (Knight et al., 1996, 1997). One missing central hole filled with a glass introduces a defect in the lattice, acting as the guiding core with the refractive index of undoped fused silica. Holes surrounding the core area serve as the cladding with the effective refractive index lower than that of the core due to presence of the air. Provided that the degree of air content expressed commonly by the k-parameter (i.e. the ratio of the air hole diameter d to the lattice pitch Λ) does not exceed 0.45 (Mortensen, 2002), the fiber supports a single transverse mode for any wavelength (the endlessly single mode regime). The most suitable manner of core expansion is scaling of cross-sectional fiber dimensions without changes in the given lattice structure. But on the understanding that the fundamental mode acts as the leaky one due to bending in the short-wavelength region, the position of bend-induced leakage boundary moves to longer wavelengths while increasing the core size (Nielsen et al., 2004b). Consequently, the spectral operation range steadily narrows that provokes fibers to be allocated on spools of greater radiuses. This prevents MOFs from being widely exploited in industrial laser or beam delivery applications with standardized curve parameters.

In this work we have concentrated our efforts on finding and implementation of a few novel MOF designs that could effectively combine the large core dimensions and the expanded spectral operation range as compared to classical MOFs. It is obvious that new structures should be actualized by applying principles different from the basic concepts of the standard MOF technology. Here we will focus on two special approaches: 1) competent manipulation of Λ-parameter in the selected wavelength region; 2) ensuring proper fiber conditions for the establishment of a substantial difference in attenuation coefficients of the fundamental (LP_{01}) and the higher order (LP_{11}) modes (differential modal attenuation).

2. MOFs with a multi-element core

2.1 Background on large-core structure design

MOF guides light along the core via the modified total internal reflection mechanism similar to that of conventional single-mode fiber (Knight, 2003). However, in some cases MOF can demonstrate specialty features that have no analogies in conventional waveguide theory.

Our previous investigations (Dukel'skii et al., 2005, 2006) indicate that the lattice pitch Λ (in general, the ratio of the wavelength λ to the pitch Λ) is the most significant parameter responsible for such main optical property as the capability of light confinement. For MOFs with air holes assembled in a triangular pattern we revealed the discrete transition between the availability of wave guidance and the lack of it. The exact position of this transition strongly depended on the Λ value.

Mentioned effect appeared in three forms: 1) absence of light canalization in short-length samples (Dukel'skii et al., 2005); 2) appreciable increase in attenuation coefficient of multimode samples with k ≥ 0.8 in the short-wavelength region of the spectra (Dukel'skii et al., 2006); 3) intensive short-wavelength leakage of the fundamental mode power into the outer fiber cladding in single-mode samples (Nielsen et al., 2004b).

For example, we observed straight fiber segments of a length from 2 to 5 centimeters in a microscope under white light launched in each sample at arbitrary angles. When Λ, defined as the hole-to-hole spacing in the first air-hole ring, exceeded ~ 10 µm, samples were characterized by the weakening of waveguiding properties in the visible part of the spectra (Dukel'skii et al., 2005). To be more precise, the core ceased to canalize light and the distribution of light over the fiber cross-section became totally uniform. No changes in the launch fiber conditions could modify that uniformity. Ultimately, the samples with $k \geq 0.8$ supported light propagation, whereas the samples with $k < 0.6$ diverged.

It is a well-known fact that conventional optical fiber made of glass materials with different refractive indexes (as well as corresponding fiber preform) can guide light irrespective of its transverse dimensions. The core diameter enlargement leads only to the increase in amount of excited modes, but not to the absolute lack of waveguiding properties. MOF technology does not imply light guidance of the initial capillary stack (preform) due to its particular structure dissimilar to the resultant MOF design. At the same time, the ability of the fiber drawn from the stack to guide light strongly depends on the transverse microstructure dimensions (d and Λ).

The next interesting phenomenon intrinsic to MOFs with a triangular cladding structure we detected after the long fiber samples (about 100 meters in length) had been investigated concerning the optical loss measurements (Figure 1).

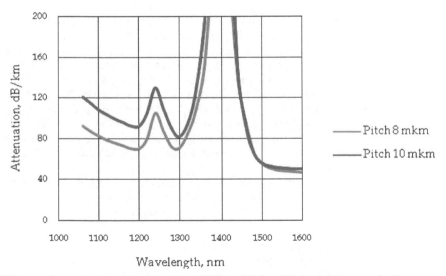

Fig. 1. Spectral attenuation pattern of the MOFs with $k \geq 0.8$ depending on the Λ-parameter (Dukel'skii et al., 2006).

As Figure 1 shows, the increase in optical losses occurred according to the fiber diameter growth, especially in the short-wavelength part of the spectra. Evidently, the Λ-parameter increased in direct proportion to the outer fiber diameter expansion. So we had the same situation as described above: the increase in hole-to-hole spacing definitely impaired wave guidance. The effect was observed while handling only multimode fiber samples with $k \geq 0.8$ and was not connected with macrobending or microbending losses.

The third aspect of Λ manipulation included the fact that abovementioned short-wavelength leakage of the fundamental mode power can dramatically enhance when core diameter rises from 20 to 35 μm (Nielsen et al., 2004b). In practical applications this unfavourable phenomenon forces MOFs to be placed on spools with augmented radiuses, extended, for example, from normal radius of 8 centimeters for communication fiber up to non-standard radius of 16 centimeters. The enhancement of optical power leakage can be formally interpreted by the impact of the Λ-parameter that proportionally increases with the fiber core diameter enlargement. So selecting suitable Λ for a given spectral region, we could control the position of modal leakage boundary.

Thus, we found out that qualified adjustment of Λ-parameter in MOFs characterized by a large core size (up to 35 μm) can be promising for the purpose of deriving a set of special properties. However, aids and concepts for achievement and implementation of these features are not trivial. One of the feasible ways to improve light confinement is to comprise the core of several elements. In this case the core becomes bigger in comparison with the core of a typical MOF structure, although the outer fiber diameter remains permanent. On the basis of geometrical considerations, it is possible to substitute not one, as ordinary, but seven or nineteen central capillaries in the initial triangular array for one solid rod made of the same material as the fiber cladding. By means of the substitution method the values of Λ-parameter can be reduced by two (7 central capillaries) or three (19 central capillaries) times in the resultant MOF as compared to a standard 1-element-core analog of the same core size.

It should be noted that there are several publications (Limpert et al., 2005, 2006) reporting on development of 7- and 19-element-core MOFs for generation of high power laser radiation, but no detailed information about modal consistence or bending performance is provided. Superior theoretical analysis (Saitoh et al., 2005) shows that the endlessly single-mode regime of operation for the 7-element-core MOF can be realized under the higher order mode cut-off condition k < 0.046. Apparently, implementation and multiple reproduction of such tiny structure correspond with huge technological difficulties. Furthermore, it seems even more difficult to carry out practically the condition for the single-mode operation of the 19-element-core MOF, which is expected to be extremely low (continuing a row for the phase higher order mode cut-off condition for the 1-element-core structure k < 0.45 and for the 7-element-core structure k < 0.046).

So taking into account the effect Λ-parameter has on the capability of light confinement, we stated a goal of creating a family of single-mode MOF structures with a large multi-element core that will not be subjected to strong influence of macrobending losses.

2.2 MOFs with the core comprised of 7 missing holes

In the first stage of our research we have successfully produced a series of MOFs with the core design presented in Figure 2. The arrangement included the 7-element core area and five air-hole rings organizing the light-reflecting cladding. All experimental samples were drawn from capillary stacks using commercially-available synthetic silica tubes and rods with OH-content in concentration of several ppm. The required value of k-parameter was obtained by the appropriate adjustment of capillary pressure in the high temperature zone of the stack during the drawing process. The outer surfaces of the elements were purified beforehand in order to reduce the influence of mechanical contaminations which content, however, in some cases was quite uncontrollable due to holding the whole technological process in normal laboratory conditions.

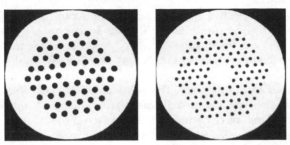

Fig. 2. Microscopic images of 1-element-core (left) and 7-element-core (right) MOF structures.

2.2.1 Core diameter of 20 μm

We started with the core size that allows a typical MOF to be single-moded over a relatively wide wavelength range in the near infrared part of the spectra. The comparison of attenuation coefficients between the 7-element-core MOF with k = 0.40 and the standard 1-element-core analog (LMA-20 produced by Crystal Fibre A/S) with k ~ 0.49 is presented in Figure 3. Optical loss measurements were made for a bending diameter of 16 centimeters.

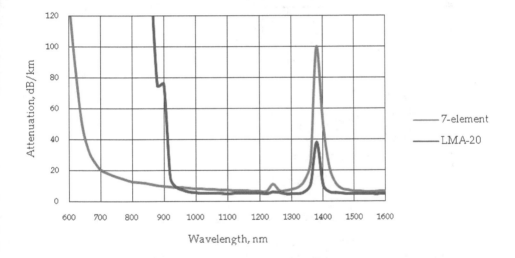

Fig. 3. Spectral attenuation pattern of 7-element-core and 1-element-core (Nielsen et al., 2004b) MOF structures.

Despite the fact that the decrease in the Λ value from 13.2 μm (Nielsen et al., 2004b) to 6 μm in our fiber has a positive effect on the position of bend-loss edge which shifts from 900 to approximately 650 nm, there is no difference in the spectral attenuation behavior between two MOF structures under study. In other words, both curves are smooth and have a low growing tendency while moving to shorter wavelengths. Moreover, they display identically the dramatic increase in attenuation coefficient due to leakage conditioned by the stationary bending radius. Two spectral peaks associated with wavelengths 1250 and 1380 nm apply naturally to hydroxyl groups absorption.

We have determined that no higher order mode cut-off can be identified. In the case of conventional single-mode fiber (for instance, SMF-28) higher order mode cut-off appears as an abrupt power decrease in attenuation curve, so far as LP_{11}-mode leaks intensively in the spectral region close to cut-off wavelength. Noted decrease can be described by the exact value of 4.8 dB (Jeunhomme, 1983) determined by either presence or absence of LP_{11}-mode radiation at the output end of the fiber. The value of 4.8 dB does not depend on the fiber length and should be observed undoubtedly in the fiber under investigation, since total light attenuation of the measured piece normally does not exceed 10 dB.

A slight decrease in the core size from 20 to 18 μm enables an additional absorption peak due to non-bonding oxygen to be clearly revealed at wavelength λ = 630 nm (Figure 4). One may notice further move of the short-wavelength bend-induced leakage boundary to the ultraviolet part of the spectra. The value of k-parameter can also affect the position of macrobending loss edge owing to a greater or lower contrast between the core and the cladding effective refractive indexes. In any case air-filling fraction needs to be controlled accurately because in some cases it may lead to multimode regime of operation.

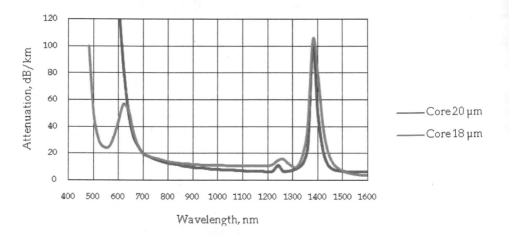

Fig. 4. Dependence of the optical losses on the decrease in the MOF core diameter.

Theoretically, the endlessly single-mode MOF is realized when k < 0.45 (Mortensen, 2002) in case the core is represented by one missing air hole. Referred fiber (Nielsen et al., 2004b) performs the lattice structure with k ~ 0.49 that is close enough to the endlessly single-mode regime condition. The fiber is definitely interpreted to be the single-mode one over the entire spectral range studied from ~ 900 to 1600 nm (Figure 3, red curve). So we suppose that the MOF with the 7-element core of 20 μm in diameter and k-parameter equal to 0.40 is also single-mode in consequence of the absence of power drops in smooth attenuation spectra from 650 to 1600 nm. Nevertheless, we have investigated roughly modal properties of the fabricated fiber. To this effect, we launched radiation from He-Ne laser (λ = 633 nm) into the sample and then observed visually typical Gaussian far-field intensity distribution on the screen placed at the distance of approximately 10 centimeters far from the fiber output. In addition, the rated value of half-divergence angle was equal to several hundredths of a radian that completely corresponded with the fundamental mode operation as well.

Those first raw results inspired us to carry more profound analysis of the 7-element-core MOF structures.

2.2.2 Core diameter of 25 μm

To verify the preceding assumption of the single-mode behavior of radiation propagated along the 7-element-core MOF with k = 0.40, extra procedures were carried out approaching the opposite to λ = 633 nm part of the spectral range. Those actions were addressed towards a new set of fibers with the core size of 25 μm in diameter and k-parameter ranging from 0.19 to 0.50.

Figure 5 shows the quasi-single-mode character of the spectral attenuation curve regardless of the k-parameter value. All presented spectra do not contain any noticeable peaks of the higher order mode cut-off. Besides that, there exists pronounced short-wavelength leakage boundary specified by the fiber placement on a spool with a diameter of 16 centimeters. One can also see that the position of this boundary depends directly on the k-parameter: the larger the k value the shorter the wavelength of bend-loss edge. Basic levels of attenuation are defined by various degree of the initial stack purification.

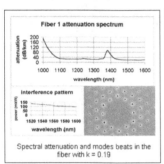
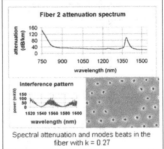
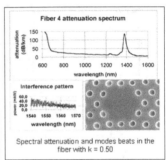

Fig. 5. Dependence of the optical losses on the MOF geometrical parameters.

The fibers, presented in Figure 5, having a length from 20 to 100 meters were investigated by means of the modal beats method. For this purpose, spectrally narrow radiation of tunable semiconductor laser (λ = 1520 – 1580 nm) was launched into a piece of SMF-28 connected with the target MOF sample. The output signal passed through a reciprocal piece of SMF-28 to PT2010 optical power meter, resulting in the modal beats pattern, registered as a change in the power distribution of light. Thus, the output signal could present the beats between two or more guided modes, i.e. intermodal interference (Figure 5, central, right), or the absolute lack of the beats in the single-mode regime (Figure 5, left). We controlled the validity of the modal beats method by comparing two schemes of the experiment in the selected spectral region. The subsidiary technique consisted in visualization of infrared radiation (λ ~ 1550 nm) on a special screen yielding an implicit coincidence with the data given by the main scanning method.

It should be noted that in spite of the smooth character of all spectral attenuation curves, given in Figure 5, and the absence of power drops corresponding to the higher order mode cut-off, the modal beats method has designated clear pattern of modal interaction. Those new results have confirmed the uncertainty of the single-mode behavior of the 7-element-core structures with the core diameter of 20 μm, examined in the previous paragraph.

In the case of excitation of the second order mode the output signal can be described according to the following expression:

$$I = A^2 + B^2 + 2AB\cos(\omega \Delta n_{eff} L / c),$$ (1)

where A and B are the amplitudes of LP_{01} and LP_{11} modes respectively, ω is the circular frequency, Δn_{eff} is the difference between effective mode indexes, L is the sample length and c is the speed of light in vacuum.

We have applied the discrete Fourier transform and, as a result, have determined the spatial frequencies v corresponding to the peaks of the interference pattern. Then we have calculated the effective index difference between the fundamental and the excited second order mode by applying formula:

$$\Delta n_{eff} = v\lambda^2 / L.$$ (2)

The experimental data are summarized in Table 1.

Fiber	k-parameter	Length, m	v, nm^{-1}	Δn_{eff}
1	0.19	19	No beats	–
1a	0.19	0.8	0.06	0.0002
2	0.27	2.0	0.8	0.00096
3	0.34	1.6	0.7	0.00105
3a	0.34	3.2	1.5	0.0010
			3.2	0.0024
4	0.50	2.8	1.3	0.0011
			1.8	0.0015
			4.0	0.0034
			4.8	0.0041

Table 1. Dependence of the effective mode index difference between the fundamental and the second order mode on the MOF structure (Agruzov et al., 2008).

As the experimental data show, vast majority of the MOF samples are characterized by the propagation of at least two guided modes. The value of the effective mode index difference $\Delta n_{eff} \sim 0.001$ exists for all presented samples regardless of the k-parameter. Only exception to the general tendency is Fiber 1 with $k \sim 0.19$, described by the absolute lack of the modal beats even when resolution capacity of registering system is the order of magnitude higher than the common level enough for the clear interference pattern visualization in all other cases. A piece of Fiber 1 having a length of 19 meters is single-moded, while a shorter piece of the same fiber (Fiber 1a) demonstrates the presence of the second order mode at the output end. So the higher order mode can be characterized by the essentially greater attenuation coefficient than the fundamental one.

Further investigations aimed at the determination of the spectral operation width of Fiber 1. But the attempt to obtain the pattern of far-field intensity distribution utilizing the available light source (He-Ne laser) failed: laser radiation intensively leaked away from the core area and filled the entire cladding even in the piece of about 1 meter in length. All other fibers from the list of Table 1 have demonstrated the modal interference pattern at $\lambda = 633$ nm: power distribution of light depended on the input fiber geometry and on the conditions of light propagation along the MOF. A bend or a mechanical stress influenced definitely the far-field pattern so the intensity peak moved from one part of the spot to another.

Reverting to the subject of the absence of LP_{11}-mode cut-off in the spectral attenuation patterns (Figures 3, 5), we should state that the higher order mode (or modes) exists simultaneously with the fundamental one in the spectral interval λ = 600 – 1600 nm. Normalized frequency V is weakly dependent on wave number. That fact differs MOFs from other types of lightguides. So for the determined wavelength range V-parameter varies negligibly (Mortensen et al., 2003), preserving its magnitude almost invariable (Figure 6). Since V-parameter directly defines the amount of guided modes, a number of them can coexist persistently within a spectral range, specific for each fiber, and at the same time undergo the infinite attenuation by reason of the huge power leakage at the identical wavelength in the blue part of the spectra.

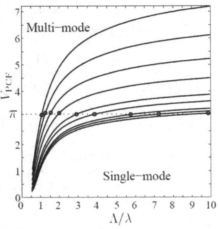

Fig. 6. Dependence of the normalized frequency V on the MOF structure (Mortensen et al., 2003). $V_{PCF} = \pi$ is the cut-off condition for the second order mode.

Finally, we have accurately determined that the single-mode operation can be achieved by the suitable selection of the k-parameter (Figure 5). The reduction of k-parameter causes the move of bend-induced leakage boundary from $\lambda \sim 650$ to $\lambda \sim 1000$ nm. Unfortunately, there is no preference of the 7-element-core MOF design over the standard 1-element-core analog (LMA-25 produced by Crystal Fibre A/S): both fibers have a bend-loss edge located in the wavelength region of 1 µm while being bent on a spool of 16 cm in diameter.

2.2.3 Core diameter of 35 µm

The improved situation takes place in case of the further core enlargement approaching 35 µm. It is a well-known fact that such great leap in the core diameter strongly affects the width of the spectral operation range (Nielsen et al., 2004b) and actually transforms the fiber into a 'single-frequency' optical element which is operable exclusively at λ = 1550 nm. We have yielded some positive results in development of the bend-resistant MOF design having a core of 35 µm in diameter.

There are three fiber samples, presented in Table 2, made of the same initial capillary stack with the air-filling fraction variable within the range k = 0.2 – 0.4. The experimental data displays the way how the reduction of k-parameter influenced the modal properties of light propagated. When k was about 0.4 the fiber was the multimode one both in red and infrared parts of the spectra. As the value of k-parameter decreased approximately twice, the fiber turned into the single-mode one, at first, in the visible part of the spectra and then in the infrared part.

Cross-section structure						
k-parameter	0.4		0.3		0.2	
Wavelength, nm	633	1550	633	1550	633	1550
Modal consistence	Multi	Multi	Single	Multi	Leakage	Single

Table 2. Dependence of the modal properties on the k-parameter reduction.

Additional investigations of the fundamental mode spot size were carried out. For this purpose, conventional single-mode fiber with the core diameter of 8 μm was attached to the MOF. Under mutual end-facet scanning the Gaussian-like power distribution of light was obtained. The MOF was excited by laser radiation at λ = 1550 nm. The results have shown that the mode spot size in the 7-element core structure is about 26 μm that corresponds equally to the 1-element core analog (Nielsen et al., 2003) despite the difference both in the amount of air holes, surrounding the core area (12 or 6), and in the k-parameter values (0.19 or ~ 0.49).

It is necessary to note that all presented MOF structures are not strongly single-mode ones if you keep in mind the existence of inherent to MOFs the endlessly single-mode regime of light propagation. For the 7-element-core microstructure this regime is provided at k < 0.046 (Saitoh et al., 2005). It is clear that the structure with such low value of k-parameter cannot be correctly realized in practice. Nevertheless, the single-mode operation can be carried out due to the significant difference in attenuation coefficients of the fundamental and the higher order modes.

For the benefit of such point of view, the behavior of the modal properties dependent on the exact k-parameter value testified (Table 2). As it may be seen, for the k = 0.3 in the red part of the spectra there existed only the fundamental mode, though in the infrared part we have observed several modes. This situation cannot be explained otherwise than by the strong attenuation coefficient of the higher order modes. In the case of conventional fiber made of the materials with different refractive indexes we observe the opposite tendency according to the expression (Snyder & Love, 1983):

$$V = 2\pi a \sqrt{n_1^2 - n_2^2} / \lambda , \qquad (3)$$

where a is the core radius, n_1 and n_2 are the core and the cladding refractive indexes respectively. The decrease of normalized frequency V causes the reduction of the amount of excited modes, so in the red part there would be several modes and in the infrared part only the fundamental one. This dependence can also be retained in the case of MOFs with correction for the effective values of the refractive index (Russell, 2006).

Finally, we declare that the higher order mode undergoes considerably strong attenuation in the visible part of the spectra than the fundamental one. This assertion has the corresponding interpretation: the shorter the wavelength the larger divergence due to diffraction and/or leakage of the higher order mode into the gaps between the air holes (Russell, 2006). By means of varying k and Λ parameters it is possible to fit the proper conditions in which the fundamental mode attenuates slightly in comparison with the higher order mode. Here we must say that the latest statement is correct only for the determined wavelength range.

Spectral attenuation pattern (Figure 7) shows the obvious preference of the 7-element-core MOF design over the standard 1-element-core analog (LMA-35 produced by Crystal Fibre A/S): the position of bend-induced leakage boundary moves to the blue part of the spectra for about 100 nm. Comparison is considered for a bending diameter of 32 centimeters.

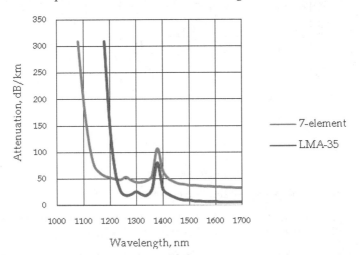

Fig. 7. Spectral attenuation pattern of 7-element-core and 1-element-core (Nielsen et al., 2004b) MOF structures.

2.3 MOFs with the core comprised of 19 missing holes

Even more impressive results have been obtained in the development of the MOFs with the core area formed by the initial substitution of 19 central capillaries in the original stack for one solid rod (Figure 8).

We have observed the similar situation for the 19-element-core MOFs with k = 0.3 as for the MOFs discussed in previous paragraph. If look carefully at the family of the spectral attenuation curves, presented in Figure 9, one can note that the increase of the core diameter leads directly to the shift of bend-induced leakage boundary of the fundamental mode to longer wavelengths within the spectral range explored. Also there is no evidence of the higher order mode cut-off in attenuation curves. All the fibers turned out to be the single-mode ones with the appreciable preference in the spectral operation range widening over the 7-element-core MOF structures and, what is more, over the 1-element-core analogs.

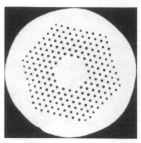

Fig. 8. Microscopic image of the 19-element-core MOF structure.

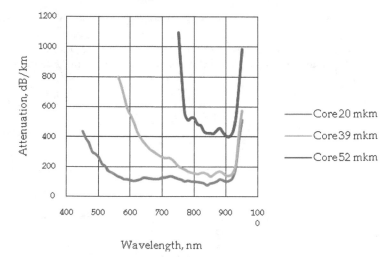

Fig. 9. Spectral attenuation pattern of the 19-element-core MOFs.

At the same time, solid analysis of the modal consistence of the 19-element-core MOFs have shown the complicated behavior of the higher order mode (Table 3).

Fiber	Core diameter, µm	Short-wavelength leakage boundary, nm (Spool diameter 16 cm)	Length, m	k-parameter	Spool diameter 16 cm		Spool diameter 32 cm	
					633 nm	1550 nm	633 nm	1550 nm
1	45	1200	6	~ 0.2	Leak.	2 modes	2 modes	
2.1	27	600	2.5	< 0.2	2 modes	2 modes		
2.2	27	600	10	< 0.2	1 mode	1 mode		
2.3	27	–	25	< 0.2	1 mode	1 mode		

Table 3. Dependence of the modal properties of the 19-element-core structure on the MOF geometrical parameters (Agruzov et al., 2010).

In fact, all presented samples are the multimode ones in case of straight fiber segments so far as the 19-element core guides a few higher-order transverse modes under k > 0.2. However, both the controllable reduction of k-parameter and the fiber placement on a spool decrease the amount of the excited modes. The single-mode behavior of the MOF bent on a spool of 16 centimeters in diameter states stationary at the fiber lengths of more than 10 meters. Then the fiber demonstrates the improved bending resistance properties and hence excellent leakage characteristic. At shorter lengths the presence of the higher order mode radiation at the output end of the fiber is unavoidable.

Particularly, Figure 10 illustrates how successfully the Λ-parameter reduction can be carried out in the 19-element-core MOF (Fiber 2.2).

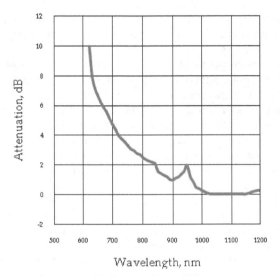

Fig. 10. Spectral attenuation pattern of Fiber 2.2 (k < 0.2).

The position of bend-loss edge is located in the red part of the spectra even when the fiber is placed on a spool of 16 centimeters in diameter. It is rather difficult to compare accurately how far the features of the 19-element-core MOF vary from the standard 1-element-core analog because of the lack of literary data. In any case the leakage characteristic is better than that of the 1-element core of 20 μm in diameter (Nielsen et al., 2004b), where optical losses increase dramatically at the wavelength $\lambda \sim 900$ nm. As in the case of the 7-element-core MOF, the single-mode propagation takes place when the k-parameter is equal or less than 0.2. For the larger k values the fiber definitely becomes the multimode one.

Our investigations (not included in Table 3) also have shown that straight pieces of Fiber 2.2 and Fiber 2.3 support several transverse modes at the lengths available in normal laboratory areas (up to 10 meters). This circumstance restricts the application potential of the 19-element-core MOFs, especially if it is necessary to obtain guaranteed single-mode regime. For example, this may take place in high power beam delivery systems, where radiation is passed from the stationary placed laser to the variable operation area. In such situation the modal consistence strongly depends on the fiber configuration. On the other hand, if one needs to transport the energy through the multibend sleeve of the fiber placed in production areas of complicated architecture, the priority of the 19-element-core MOF is evident.

2.4 Special attenuation mechanism

While investigating the modal properties of the aforementioned MOF structures with the large 7- or 19-element core, we have defined a specific mechanism when the attenuation coefficient of the higher order mode substantially exceeds the same parameter of the fundamental one. This phenomenon had the most decisive effect on the modal consistence of the MOFs with the core of 35 μm in diameter, allowing the fibers with k < 0.2, being bent on a standard spool with a bending diameter of 16 centimeters, to propagate only the fundamental mode within a broad spectral range λ = 600 – 1550 nm. Draw attention that, theoretically, the higher order mode cut-off condition for this fiber is expected to be technologically unfeasible (since for the 7-element-core fiber k-parameter, in theory, has to be as low as 0.046 to ensure the single-mode operation). So the great difference in attenuation coefficients of LP_{01} and LP_{11} modes enables the implementation of the 19-element-core MOF with the absolutely workable k ~ 0.2, operating in the single-mode regime (Fiber 2.2 and Fiber 2.3).

To determine the proportion of the attenuation coefficients, sufficient for the single-mode operation via the differential modal attenuation, we have estimated the optical losses of the higher order mode. For this purpose, we have measured the depth of modulation while registering several patterns of the modal beats. The length of the investigated Fiber 2.2 sample was varied deliberately to achieve distinct patterns (1 meter, 1.5 meters and 2 meters). This procedure has shown the inverse influence of the fiber length on the depth of modulation: less pronounced pattern corresponded to larger sample lengths. Thus, the attenuation coefficient of the higher order mode was evaluated to be ~ 5 dB/m. At the same time, the fundamental mode attenuation coefficient, measured by a cut-back technique, have been rated at a level of tens of dB/km. So now we can state that the fundamental mode propagation via the differential modal can be effectively realized in MOF structures when the higher order mode attenuation coefficient is of at least two orders of magnitude larger than the analogous parameter of the fundamental mode.

The next goal consisted in detailed analysis of the conditions opportune enough for the establishment of the differential modal attenuation. The implementation of microstructures with large cores, irrelevant to 7 or 19 elements, and great Λ-parameter values seemed to be the most applicable means to collect the data.

3. MOFs based on the differential modal attenuation mechanism

In the previous part of the work the special attenuation mechanism has been shown. It described the situation when the attenuation coefficient of the fundamental mode may be essentially lower than the same optical parameter of the higher order mode. Starting from that point, we have concentrated our efforts on the extensive research of a few novel MOF designs that could successfully correspond with a set of special requirements: appreciable difference in optical losses of the first two modes (LP_{01} and LP_{11}), single-mode operation, high bending resistance and fiber placement on a standard transport spool of 16 centimeters in diameter. Among the others we have tested structures with the circular cladding distribution, with the special C_{3V} cladding symmetry and with the 1-element shifted core. All of these MOF structures seemed to fit with the requirements, especially with the enforcement of the higher order mode to undergo the enhanced attenuation. We can say in advance that the differential modal attenuation mechanism better displays in case of the core diameters of more than 30 μm and the k-parameters as large as 0.60.

3.1 Investigation procedures

In order to investigate carefully the modal properties of the MOFs as a function of the transverse fiber dimensions, we used the layout: a set of semiconductor lasers (λ = 658, 808, 980 and 1550 nm), objective lenses, micrometer screws, digital CCD-camera (640 x 480 with the pixel size of 7 μm), fiber cutter and personal computer for data handling. We have been registering near-field or far-field patterns of the radiation propagated along the fibers while varying the input coupling conditions. For this purpose, laser radiation was launched into the test fiber under diverse apertures to achieve the most powerful signal on a CCD-camera screen. The fundamental mode specified by the Gaussian power distribution of light was the easiest to excite. If under varying the input coupling geometry we observed only the fundamental mode alteration (the modal spot became larger or smaller uniformly) and the higher order mode did not appear at the output end of the fiber, we explicitly considered the situation to be the single-mode one (Figure 11, left). The amount of modes in tables below was noted as 1. In the other case, when we clearly ascertained the distortion of the Gaussian power distribution or mode superposition with nearly equivalent peak power levels (Figure 11, right), we denoted the amount of the excited modes as 2.

In addition, we have made the evaluation of the mode spot size of the fiber samples with the definite single-mode regime of operation. We have modified the well-known expression for the half-divergence angle (Mortensen et al., 2002) to the form:

$$\omega = \lambda L / \pi W , \qquad (4)$$

where W is the mode spot size measured at $1/e^2$ level of peak intensity on the CCD-camera screen and L is the distance between the screen and the fiber end-facet.

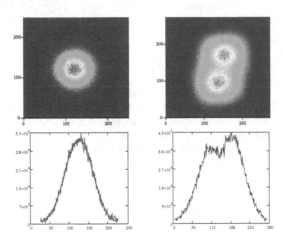

Fig. 11. Far-field patterns of the MOFs: fundamental mode propagation with the Gaussian approximation (left) and mode superposition (right).

3.2 MOFs with the circular cladding distribution

We have designed and manufactured a novel type of silica-based MOF containing a solid glass core of 30 μm in diameter and three air-hole rings organizing light-reflecting cladding, as it is shown in Figure 12. The similar structure containing 8 circles was mentioned in (Martelli et al., 2007)

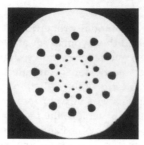

Fig. 12. The MOF structure with the circular cladding distribution.

The principal difference between the standard triangular MOF structure and the circular one is that in the latter case the amount of air-filled channels does not increase within next ring but remains invariable while going from the core area to the outer fiber boundary. In whole, the number of air holes in each ring has to be relatively large to provide a great contrast between the core refractive index and the effective cladding refractive index. In our case the number of air holes in each of the successive rings surrounding the core area was 12, which, in our opinion, is large enough to guarantee satisfactory bending characteristics. On the other hand, a quite large number of air holes will ensure good light confinement in the core area. Another important feature, worthy of attention, is the shape of the fundamental mode spot. So far as the air holes distribution in the first ring replicates the form of nearly a circle, the MOF structure performs the circular-like shape of the modal spot (C_{12v} symmetry), that in some cases may be more preferable as compared to a typical six-fold rotational one (C_{6v} symmetry).

As it is illustrated in Figure 12, at first, we have tended to equalize the k-parameter value in the air-hole rings, because in the triangular lattice that parameter, actually, is approximately constant over the MOF cross-section. In order to fabricate this structure we used three sets of the initial capillaries made of quartz glass. They were characterized by the outer diameters of 1.25, 2.10 and 3.30 millimeters. The ratio of the inner to the outer diameters in all three sets of capillaries was equal to ~ 0.50.

We have managed to achieve the single-mode operation over the entire spectral range studied (λ = 658 – 1550 nm) in the derived structure, though the attenuation coefficient was too large in consequence of the tremendous leakage of the fundamental mode power into the fiber curve (standard spool of 16 centimeters in diameter). The weak dependence of the mode spot diameter on the wavelength confirms that process (Figure 13).

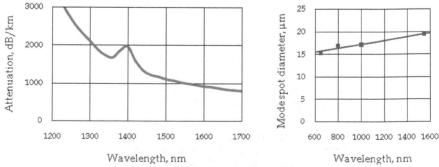

Fig. 13. Spectral attenuation pattern (left) and mode spot diameter (right) of the first series of the MOFs with the circular cladding distribution.

The sensitivity of the propagated radiation to bend may be explained largely by its great leakage through the silica gaps between the air holes in the external ring.

In the second and third series of MOFs with the circular cladding distribution we deliberately used the initial capillaries with the increased ratio of the inner to the outer diameters (Table 4).

Air-hole ring	Ratio in the second series	Ratio in the third series
1 (nearest to the core area)	0.8	0.8
2	0.75	0.8
3	0.75	0.8

Table 4. Inner/outer diameters ratio for the circular-cladding structures.

Using the capillaries with the geometrical parameters presented in Table 4 we have obtained tangible reduction in the optical losses of the fundamental mode. Moreover, there was no evidence of the second order mode radiation both in 3-meter-length and 30-meter-length samples. The leakage losses of the LP_{11} mode have been estimated to be about 10 dB/m. The experimental results are accumulated in Table 5.

Core diameter, μm	Modal consistence (Fundamental mode spot diameter, μm)				Cross-section
	$\lambda = 658$ nm	$\lambda = 808$ nm	$\lambda = 980$ nm	$\lambda = 1550$ nm	
30	1 (17.1)	1 (18.3)	1 (18.8)	1 (24.0)	
40	Leakage	1 (25.7)	1 (27.3)	1 (31.5)	

Table 5. Modal consistence of the second (30 μm) and the third (40 μm) series of the MOFs with the circular cladding distribution.

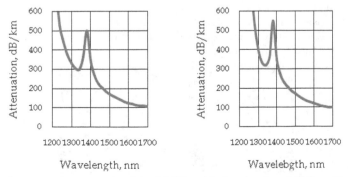

Fig. 14. Spectral attenuation pattern of the MOFs with the circular cladding distribution: the core diameters are 30 μm (left) and 40 μm (right).

As Figure 14 illustrates, the MOF with the core diameter of 30 μm has excessively large optical losses. However, a certain increase in the air-filling fraction in each of three air-hole rings and simultaneous enlargement of the core size (up to 40 μm) give the structure optimal correlation between high attenuation coefficient of the upper modes and reasonable bend performance of the fundamental mode. Table 5 shows that the MOF with the core diameter of 40 μm is characterized by the relatively large mode spot size, ranging from 25.7 to 31.5 μm in the inspected spectral range. Despite the significant attenuation coefficient, about 0.15 dB/m at λ = 1550 nm, the MOF is operable for a standard bending diameter of 16 centimeters. In identical conditions, a well-known triangular structure with the 1-element core undergoes the infinite attenuation (Nielsen et al., 2004b).

We assert that the considerable reduction in the attenuation coefficient of the MOFs with the circular cladding distribution can be achieved by both proper variation of the air-hole sizes and appropriate adjustment of the silica gaps between them. Moreover, the presence or the absence of the higher order mode at the output end of the fiber is determined mainly by the air content in the first air-hole ring. At the same time, the air-filling fractions of the second and the third rings, especially the external, define the leakage loss level of the fundamental mode. As additional experiment has shown, the minimal attenuation in the multimode regime (Figure 15) was measured to be ~ 8 dB/km caused mainly by impurities. This MOF has a potential to be used in high-NA applications.

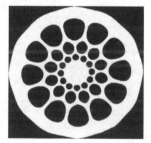

Fig. 15. Microscopic image of the multimode MOF structure with the circular cladding distribution.

3.3 MOFs with the C₃ᵥ cladding symmetry

We have also fabricated a series of the MOF structures that can be treated as an adaptation of standard 1-element-core design with a triangular arrangement of air holes (Russell, 2006). A key variation consisted in the intentional introduction of an alternation of large and small air holes to the MOF design attaining so called C_{3v} transverse symmetry (Figure 16), as opposed to the standard C_{6v} symmetry.

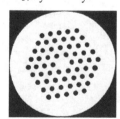 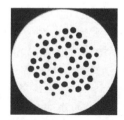

Fig. 16. Microscopic images of the MOFs: typical C_{6v} symmetry (left) and non-standard C_{3v} symmetry.

The alternation of large and small air holes, especially in the air-hole ring closest to the core area, may confidently lead to the enhancement of the higher order mode attenuation. The main mechanism remained the same we had applied earlier for the structure with the circular cladding distribution. We talk about the controllable leakage of the second order mode power through the silica gaps between each pair of neighboring holes.

To achieve the single-mode operation in presented MOF structures we used quartz capillaries of the same outer diameter (1.75 μm), but of the diverse inner diameter. The exact ratios of the inner/outer diameters were 0.6 and 0.8. Additionally, we controlled the air-filling fraction during the fiber drawing process by varying two essential technological parameters we could easily change. We mean the capillary pressure and the drawing temperature. The experimental results are given in Table 6. Optical measurements were made for a bending diameter of 16 centimeters.

Core diameter, μm	Modal consistence (Fundamental mode spot diameter, μm)				Cross-section
	λ = 658 nm	λ = 808 nm	λ = 980 nm	λ = 1550 nm	
25	Leakage	Leakage	1 (16.8)	1 (19.6)	
35	Leakage	Leakage	Leakage	1 (24.4)	

Table 6. Modal consistence of the MOFs with the C$_{3v}$ cladding symmetry.

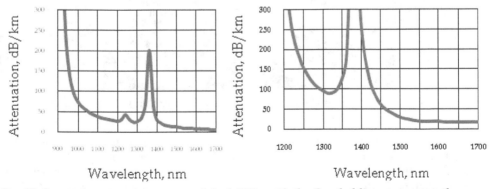

Fig. 17. Spectral attenuation pattern of the MOFs with the C$_{3v}$ cladding symmetry: the core diameters are 25 μm (left) and 35 μm (right).

We have detected some positive features of the MOFs with the C$_{3v}$ cladding symmetry including single-mode regime and high bending resistance properties. The most impressive situation takes place in the fiber with the core diameter of 35 μm, as the spectral operation range can be expanded for about 300 nm to the blue part of the spectra in comparison with the MOF of a typical triangular configuration with the air holes of invariable dimensions. Classical MOFs are still required to be placed on non-standard spools having a diameter of 32 centimeters (Nielsen et al., 2004b), while our fibers are operable for a bending diameter of 16 centimeters. In addition, as opposed to the MOFs with the circular cladding distribution (discussed in previous paragraph), the fibers with the C_{3v} cladding symmetry can guarantee optical losses of less than 20 dB/km at λ = 1550 nm (Figure 17).

3.4 MOFs with the shifted core

This paragraph will cover in detail main optical properties as well as bend characteristics of another MOF structure based on the declared concept of the differential modal attenuation. Theoretically, the higher order mode of a typical MOF with triangular arrangement of air holes, as opposed to the fundamental mode, may be characterized by a significantly larger degree of the power penetration into the fiber cladding (Tsuchida et al., 2005; Russell, 2006). Applying this specific mode feature, we made an assumption that the MOF structure with the core shifted for the pitch value from its usual location in the center of the lattice structure will exhibit enhanced leakage losses of the higher order mode. At the same time, the leakage losses of the fundamental mode will maintain their value as if the fiber structure is completely retained. So we can implement the special MOF design that enables a relatively simple control of the higher order mode attenuation. A conditioning factor, as in the previous cases we have already studied (MOF structures with the circular cladding distribution and the C_{3v} cladding symmetry), is the reasonable balance towards the dimensions of the air holes, especially in the ring closest to the core area, and the spaces between them filled with a glass material. We have successfully produced by classical stack-and-draw technique the following MOF structure (Figure 18).

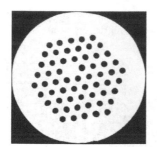 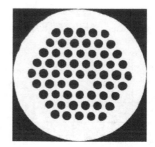

Fig. 18. Microscopic images of the fabricated MOF structures with the shifted 1-element core: k = 0.45 (left), k = 0.75 (right).

When the core is displaced from the center of the lattice, the peripheral part of the effective LP_{11}-mode area may intensively leak away into the outer fiber cladding which is then situated nearer than in the MOF of a standard central configuration. Thus, the fundamental mode operation is not strongly defined by the phase cut-off condition, known as k < 0.45 for the second order mode (Mortensen, 2002). As we will show further, the endlessly single-mode regime may be carried out at certain lengths of the fiber even when the k-parameter value is for more than 40% larger relative to its theoretical value. In this case the Λ-parameter and the curve diameter resulting in the bend performance are crucial for the establishment of the differential mode attenuation.

The experimental results are collected in Table 7. It should be noted that we intentionally fabricated several samples with the standard central-core design to compare two different transverse structures. All optical measurements were made for a bending diameter of 16 centimeters.

One can see from Table 7 that the core diameter enlargement from 12.5 to 20 μm under the k-parameter being nearly constant (k ~ 0.50) caused the additive attenuation of the higher order mode (samples 2_{SHIFT} and 3_{SHIFT}).

Fiber	Core diameter, μm	k	Length, m	Modal consistence at λ (Fundamental mode spot size, μm)			
				658 nm	808 nm	980 nm	1550 nm
1_{SHIFT}	12.5	0.35	10	1 (10.4)	1 (11.8)	1 (11.8)	1
2_{SHIFT}	12.5	0.52	10	2	2	2	2
3_{SHIFT}	20.0	0.51	10	1 (13.4)	1 (14.1)	1 (14.2)	1
4_{SHIFT}	20.0	0.60	10	1 (12.3)	1 (13.0)	1 (13.4)	1
5_{SHIFT}	20.0	0.81	10	2	2	2	> 2
1_{CENT}	20.0	0.65	20	Leakage	1 (15.3)	1 (15.8)	2
2_{CENT}	22.0	0.65	20	Leakage	1 (13.9)	1 (14.9)	2
3_{CENT}	22.0	0.67	20	1 (12.6)	1 (13.7)	1 (14.0)	2

Table 7. Modal properties of the MOFs with the shifted and the central cores.

In theory, the core expansion leads to the increase of normalized frequency V, representing exactly the amount of guided modes. This situation can be simply illustrated with the expression (Birks et al., 1997):

$$V = 2\pi\Lambda\sqrt{n_{co}^2 - n_{cl}^2} / \lambda,$$ (5)

where n_{co} and n_{cl} are the refractive index of the core and the effective refractive index of the cladding respectively.

So when V-parameter increases, the existing modes become less sensitive to perturbations, for example, to macro- and microbending, and the higher order mode attenuation coefficient has to be lower. The alternative process in our experiment may be explained by the mechanism of the enhanced modal leakage through the spaces between the air holes, which dimensions extend in the direct proportion to Λ-growth when the whole structure is scaled. Generally, the modal consistence of the MOFs with shifted and central cores approximately coincides if deal with the same core diameters and k-parameters values. The most impressive difference is that in the visible part of the spectra (λ = 658 nm) the fundamental mode radiation of the central-core MOFs intensively leaks away from the core area into the outer fiber cladding, while it good confines in the shifted-core ones. Such high bending resistance properties can be elucidated by the increased air-filling fraction (sample 4_{SHIFT}) and the mode spot size that is decreased as compared to the spot size in the central-core MOF (samples 4_{SHIFT} and 1_{CENT}). In addition, the mode spot size in MOFs with the shifted core is weakly dependent on the wavelength that is a distinctive feature of MOFs as a class of lightguides with the triangular arrangement of the air holes (Nielsen et al., 2004a). Despite the slight reduction in the fundamental mode spot size, the shifted-core MOFs have an obvious preference over the central-core analog. It consists in the expansion of the spectral operation range.

Thus, we have accurately determined that the position of bend-induced leakage boundary of the MOFs with the shifted core is located in the visible part of the spectra. To identify it numerically we have produced several samples of the length from 50 to 100 meters. The respective curve is presented in Figure 19.

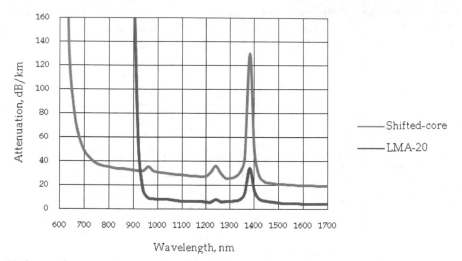

Fig. 19. Spectral attenuation pattern of the MOFs with the core diameter of 20 μm.

As Figure 19 shows, leakage of the fundamental mode power occurs in the wavelength region λ ~ 650 nm. At the same time, in the central-core MOF with the identical core size strong leakage occurs at λ ~ 900 nm (spool diameter of 16 centimeters) and at λ ~ 650 nm (spool diameter of 32 centimeters) (Nielsen et al., 2004b). In other words, the MOF with the shifted core can be characterized by the noticeably improved stationary bending resistance due to the relatively large value of k-parameter which is equal to 0.60. The spectral operation range widening is about 250 nm that is an obvious preference over the typical MOF structure.

Similar processes can be observed in the MOF with the shifted core of 34 μm in diameter (Figure 20).

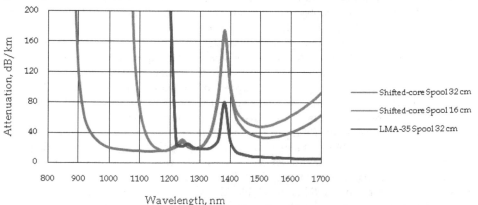

Fig. 20. Spectral attenuation pattern of the MOFs with the core diameter of 34 - 35 μm.

The increase in the k-parameter value up to 0.65 allows the fiber being placed on a standard spool of 16 centimeters in diameter to operate still in the single-mode regime. The mode spot size is approximately 26 μm that coincides with the data for the central-core MOFs (Nielsen et al., 2003).

We assert that further increase in the air-filling fraction (up to 0.70), caused by the natural desire to expand the range of working frequencies as much as possible, leads to the display of the impurity of the higher order mode in the power distribution of light at the output end of the fiber. So we consider the MOF with the shifted core of 34 – 35 μm in diameter and the k-parameter of 0.65 to be the optimal bend-resistant design combining robust fundamental mode propagation and the expanded spectral operation interval.

It should be noted that the investigated MOF with the shifted core of 34 μm in diameter exhibits a certain growth of the attenuation coefficient in the spectral region λ > 1200 nm (Figure 20). As we have already ascertained mode spot size is weakly dependent on the working wavelength in the fibers with the air holes assembled in the triangular array (Nielsen et al., 2004a) and the fundamental mode leakage occurs in the short-wavelength part of the spectra (Nielsen et al., 2004b). When the fiber core is shifted, the effective mode area largely penetrates into the outer fiber cladding as compared to the central-core fiber, that corresponds adequately with the certain increase in the attenuation coefficient while going to longer wavelengths.

The validity of the ultimate role of differential modal attenuation in achievement of the single-mode operation may be illustrated by the following data (Table 8).

Length, m	Wavelength, nm			
	λ = 658 nm	λ = 808 nm	λ = 980 nm	λ = 1550 nm
MOFs with the shifted core and k = 0.65				
50	1 mode	1 mode	2 modes	
5	1 mode	1 mode		
3	1 mode	2 modes		
1.5	2 modes	2 modes		
MOFs with the shifted core and k =0.68				
40	1 mode	1 mode	2 modes	
5	1 mode	2 mode		
2.5	2 mode	2 mode		

Table 8. Dependence of the modal consistence on the fiber length.

Table 8 illustrates the modal consistence of the short-length (from 1.5 to 5 meters) samples of the MOFs with the shifted core of 20 μm in diameter. We have compared these experimental data with those obtained for the long-length samples. During the measurement procedures the fibers were placed on standard spools of 16 centimeters in diameter. As one can see, short-length samples were characterized by the multimode regime of operation which transformed into the single-mode while increasing the fiber length. The higher order mode attenuated almost completely after passing the length of approximately 5 meters (sample with k = 0.65). On the basis of the statement that total attenuation of the higher order mode could be estimated to be of at least 10 dB, we got the bottom boundary for the attenuation coefficient equal to ~ 2 dB/m at λ = 658 - 808 nm. That result corresponded with the data we have obtained earlier while working on the MOFs with a multi-element core. Table 8 shows that further increase in the k-parameter up to 0.68 decreased the attenuation coefficient of the higher order mode that, in turn, limited steadily the spectral operation range.

Finally, we declare that in optimized conditions the expansion of the spectral operation range of the shifted-core MOFs may achieve 300 nm in the near infrared part of the spectra (Figure 20) as compared to the central-core MOFs (Nielsen et al., 2004b). This effectively combines with the large air-filling fraction (k = 0.65) that makes the fiber less susceptible to macrobending or microbending effects.

4. Conclusion

In this work, we have reported on the recent results in investigating main optical properties of several series of MOFs. Two of them (7- and 19-element-core designs) have already been discussed earlier, whereas three others (the circular cladding distribution, the C_{3v} cladding symmetry and the shifted-core structure) represent novel design. The modal consistence of radiation propagated along the fiber cores and the spectral attenuation curves have been provided.

We have accurately determined that the basic priorities of novel MOF structures are the expanded spectral operation range (up to 300 nm) and the bend-resistant performance. We have managed to achieve these features by ensuring proper conditions for the high air-filling fraction of the cladding structure. Simultaneously, we have successfully assured the condition for the higher order mode to undergo strong attenuation via considerable power leakage into the outer fiber cladding. That point principally diverges our approach from the classical one, which guarantees the phase cut-off condition for the propagation of the second order mode. The mechanism of the differential modal attenuation has been clearly stated in the case of the shifted-core MOF design when the core area was located close to the outer fiber cladding. The mentioned circumstance has led to the strong attenuation coefficient of the higher order mode that has been estimated to be in the vicinity of 2 - 6 dB/km. In other cases the physical foundation of the differential modal attenuation was not so clear, though it effectively worked allowing the MOFs to support practically only a single transverse mode over a broad spectral range.

The fiber designs discussed in this work can be successfully applied in high power laser technology or in laser beam delivery applications with standardized curve parameters.

5. References

Agruzov, P. M.; Kozlov, A. S.; Petrov, M. P.; Dukel'skii, K. V.; Komarov, A. V.; Ter-Nersesyants, E. V.; Khokhlov, A. V. & Shevandin, V. S. (2008). Mode composition of holey fibers with a large seven-element core. *Journal of Optical Technology*, Vol. 75, No. 11, (November 2008), pp. 747-749, ISSN 1070-9762

Agruzov, P. M.; Dukel'skii, K. V.; Komarov, A. V.; Ter-Nersesyants, E. V.; Khokhlov, A. V. & Shevandin, V. S. (2010). Developing microstructured lightguides with a large core, and an investigation of their optical properties. *Journal of Optical Technology*, Vol. 77, No. 1, (January 2010), pp. 59-62, ISSN 1070-9762

Birks, T. A.; Knight, J. C. & Russell, P. St. J. (1997). Endlessly single-mode photonic crystal fiber. *Optics Letters*, Vol. 22, No. 13, (July 1997), pp. 961-963, ISSN 0146-9592

Dukel'skii, K. V.; Komarov, A. V.; Ter-Nersesyants, E. V.; Khokhlov, A. V. & Shevandin, V. S. (2005). Realization of photonic crystal fibers in S.I. Vavilov Federal Optical Institute. *Proceedings of AIS'05 and CAD'2005*, ISBN 5-9221-0621-X, Divnomorsk, September 2005

Dukel'skii, K. V.; Kondrat'ev, Yu. N.; Komarov, A. V.; Ter-Nersesyants, E. V.; Khokhlov, A. V. & Shevandin, V. S. (2006). How the pitch of a holey optical fiber affects its lightguide properties. *Journal of Optical Technology*, Vol. 73, No. 11, (November 2006), pp. 808-811, ISSN 1070-9762

Jeunhomme, L. B. (1983). *Single-Mode Fiber Optics: Principles and applications*. Marcel Dekker, ISBN 978-082-4770-20-4, New York, USA

Knight, J. C.; Birks, T. A.; Russell, P. St. J. & Atkin, D. M. (1996). All-silica single-mode optical fiber with photonic crystal cladding. *Optics Letters*, Vol. 21, No. 19, (October 1996), pp. 1547-1549, ISSN 0146-9592

Knight, J. C.; Birks, T. A.; Russell, P. St. J. & Atkin, D. M. (1997). All-silica single-mode optical fiber with photonic crystal cladding: errata. *Optics Letters*, Vol. 22, No. 7, (April 1997), pp. 484-485, ISSN 0146-9592

Knight, J. C. (2003). Photonic crystal fibres. *Nature*, Vol. 424, No. 6950, (August 2003), pp. 847-851, ISSN 0028-0836

Limpert, J.; Deguil-Robin, N.; Manek-Honninger, I.; Salin, F.; Roser, F.; Liem, A.; Schreiber, T.; Nolte, S.; Zellmer, H.; Tunnermann, A.; Broeng, J.; Petersson, A. & Jakobsen, C. (2005). High-power rod-type photonic crystal fiber laser. *Optics Express*, Vol. 13, No. 4, (February 2005), pp. 1055-1058, ISSN 1094-4087

Limpert, J.; Schmidt, O.; Rothhardt, J.; Roser, F.; Schreiber, T.; Tunnermann, A.; Ermeneux, S.; Yvernault, P. & Salin, F. (2006). Extended single-mode photonic crystal fiber lasers, *Optics Express*, Vol. 14, No. 7, (April 2006), pp. 2715-2720, ISSN 1094-4087

Martelli C.; Canning J.; Gibson B. & Huntington S. *Optics Express*, Vol. 15, No. 26, (December 2007), pp. 17639-17644, ISSN 1094-4087

Mortensen, N. A. (2002). Effective area of photonic crystal fibers. *Optics Express*, Vol. 10, No. 7, (April 2002), pp. 341-348, ISSN 1094-4087

Mortensen, N. A.; Folkenberg, J. R., Skovgaard, P. M. W. & Broeng, J. (2002). Numerical aperture of single-mode photonic crystal fibers. *IEEE Photonic Technology Letters*, Vol. 14, No. 8, (August 2002), pp. 1094-1096, ISSN 1041-1135

Mortensen, N. A.; Folkenberg, J. R.; Nielsen, M. D. & Hansen, K. P. (2003). Modal cut-off and the V-parameter in photonic crystal fibers. *Optics letters*, Vol. 28, No. 20, (October 2003), pp. 1879-1881, ISSN 0146-9592

Nielsen, M. D.; Folkenberg, J. R. & Mortensen, N. A. (2003). Single-mode photonic crystal fiber with an effective area of 600 μm^2 and low bending losses. *Electronics Letters*, Vol. 39, No. 25, (December 2003), pp. 1802-1803, ISSN 0013-5194

Nielsen, M. D.; Folkenberg, J. R.; Mortensen, N. A. & Bjarklev, A. (2004). Bandwidth comparison of photonic crystal fibers and conventional single-mode fibers. *Optics Express*, Vol. 12, No. 3, (February 2004), pp. 430-435, ISSN 1094-4087

Nielsen, M. D.; Mortensen, N. A.; Albertsen, M.; Folkenberg, J.R.; Bjarklev, A. & Bonacinni, D. (2004). Predicting macrobending loss for large-mode area photonic crystal fibers. *Optics Express*, Vol. 12, No. 8, (April 2004), pp. 1775-1779, ISNN 1094-4087

Russell, P. St. J. (2006). Photonic-Crystal Fibers. *Journal of Lightwave Technology*, Vol. 24, No. 12, (December 2006), pp. 4729-4749, ISSN 0733-8724

Saitoh, K.; Tsuchida, Y.; Koshiba, M. & Mortensen, N. A. (2005). Endlessly single-mode holey fibers: the influence of core design. *Optics Express*, Vol. 13, No. 26, (December 2005), pp. 10833-10839, ISSN 1094-4087

Snyder, A. W. & Love, J. D. (1983). *Optical Waveguide Theory*. Chapman and Hall Ltd, ISBN 978-041-2099-50-2, London, UK

Tsuchida, Y.; Saitoh, K. & Koshiba, M. (2005). Design and characterization of single-mode holey fibers with low bending losses. *Optics Express*, Vol. 13, No. 12, (June 2005), pp. 4770-4779, ISSN 1094-4087

Tunnermann, A.; Schreiber, T.; Roser, F.; Liem, A.; Hofer, S.; Zellmer, H.; Nolte, S. & Limpert, J. (2005). The renaissance and bright future of fibre lasers. *Journal of Physics B: Atomic, Molecular and Optical Physics*, Vol. 38, No. 9, (April 2005), pp. S681-S693, ISSN 0953-4075

4

Influence of Current Pulse Shape on Directly Modulated Systems Using Positive and Negative Dispersion Fibers

Paloma R. Horche[1] and Carmina del Río Campos[2]
[1]ETSI Telecomunicación, Universidad Politécnica de Madrid,
[2]Escuela Politécnica Superior, Universidad San Pablo CEU,
Spain

1. Introduction

The proliferation of high-bandwidth applications has given rise to a growing interest, between network providers, on upgrading networks to deliver broadband services to homes and small businesses. There has to be a good balance between the total cost of the infrastructures and the services that can be offered to the end users, as they are very sensitive to equipment outlay, requiring the use of low-cost optical components. Coarse Wavelength Division Multiplexing (CWDM) is an ideal solution to the tradeoff between cost and capacity (Thiele, 2007). This technology uses all or part of the 1270 to 1610 nm wavelength fiber range with an optical channel separation of about 20 nm. This channel separation allows the use of low-cost, uncooled, Directly Modulated Lasers (DML). The main advantages of these devices associated with uncooled operation are (Nebeling, 2007); No integration of TEC and cooler required, less complexity for control electronics, reduced power consumption, only laser diode current required, lower device cost. Otherwise, the direct modulation of the laser current leads to a modulation of the carrier density giving rise to a *chirp frequency*. This results in a broadened linewidth and a laser wavelength drift (Henry, 1982; Linke, 1985). Since wavelength chirp was recognized several years ago, (Koch, 1988; Hinton, 1993) many papers have addressed the causes and implications of chirp on the optical system performance (Cartledge, 1989; Hakki, 1992; Horche, 2008). However, a large part of the research has focused on fiber transmission properties. The idea of these studies is, generally, based on considerations regarding the interaction of the laser frequency chirp with the fiber dispersion. It is known that if the chirp parameter (α-factor) is positive, as is always the case for directly modulated lasers, then the frequency components of the leading edge of the pulse will be blue-shifted and the trailing edge red-shifted. If the pulses are transmitted over a negative dispersion fiber where the blue wavelength is slower than the red, to some extent, pulse compression and significant transmission performance improvement is expected (Morgado & Cartaxo, 2001; Tomkos et al., 2001).

To counteract the effect of chromatic dispersion, the system can be moved to a non-zero negative dispersion-shifted fiber, but this method has some problems that are difficult to resolve when it comes to a CWDM metropolitan or access network; The traffic must be interrupted. This method may only be used for upgrading long wavelength channels where this fiber has much lower dispersion values. And, cost-effectiveness, changing the fiber is expensive both to buy and to lay.

In DMLs, the frequency chirp for large-signal modulation can be determined from the shape of the electric pulse applied to the optical source (modulation current) (Coldren & Corzini, 1995). Since the DML chirp parameter is tunable through the modulation current applied to the diode laser, optimum operating-conditions in terms of the chirp/dispersion interactions can be set for fibers having different amounts and signs of dispersion.

It is the aim of this Chapter to discuss and compare how the shape of the modulated signal (e.g, exponential-wave, sine-wave, Gaussian, etc.) can improve the system performance when using both positive and negative dispersion fibers. With this method it is possible to improve each of the WDM system channels individually, offering a low-cost solution since it only involves changes in the transmitters and avoiding the replacement of the fiber.

The aim of this chapter is also to present analytical and simulation results pertaining to the transmission of chirped optical signals in a dispersive fiber.

This Chapter is organized as follows: Section 2 deals with the theoretical background of optical fiber dispersion; In Section 3 we analyze the power and chirp waveforms for different DMLs both theoretically and by simulation; Sections 4 and 5 focus on the study of the interplay between the laser chirp and the dispersion of both Gaussian pulses and pulses of an arbitrary shape, upon which the subject of this work is based; Section 6 details the system modeling together with the presentation of the parameters used in our simulation for the different fibers, DMLs and modulation current; Section 7 discusses the simulation results for the determination of the optimum modulation current under different scenarios; Section 8 studies the accumulated phase along the link as a method of improving the system performance. In Section 9 we analysis the transmission performance over different lengths of positive and negative dispersion fibers through computer simulation. Finally, a brief summary and the conclusions are presented in Section 10.

2. Factors contributing to dispersion in optical fibers

When a short pulse of light travels through an optical fiber, its power is "dispersed" in time so that the pulse spreads into a wider time interval. The principal sources of dispersion in single optical fibers are material dispersion, waveguide dispersion, polarization-mode dispersion and non-linear dispersion. The combined contributions of these effects to the spread of pulses in time are not necessarily additive (Agrawal, 2010).

The PMD-induced pulse broadening is characterized by the root mean-square (RMS) $\sigma_T \equiv D_p\sqrt{L}$, where D_p is the PMD parameter. Measured values of D_p vary from fiber to fiber in the range $D_p = 0.01–1$ ps/$\sqrt{km}$. Because of the $\sqrt{L}$ dependence, PMD induced pulse broadening is relatively small compared with the other effects. Indeed, for fiber lengths < 400 km, as is the case of metro-CWDM systems, the PMD can be ignored.

The combined effects of material dispersion and waveguide dispersion which is referred to as chromatic dispersion is, in general, the major source of pulse broadening and it may be determined by including the wavelength dependence of the optical fiber refractive indices, n_1 and n_2 when determining $d\beta/d\omega$ from the characteristic equation of the fiber.

Since β is a slowly varying function of this angular frequency, one can see where various dispersion effects arise by expanding β in a Taylor series about a central frequency ω_0 (Keiser, 2010). Expanding β in a Taylor series yields

$$\beta(\omega)=\beta_0+\beta_1(\omega-\omega_0)+\frac{1}{2}\beta_2(\omega-\omega_0)^2+\frac{1}{6}\beta_3(\omega-\omega_0)^3+\cdots \tag{1}$$

where β_m denotes the mth derivative of β with respect to ω evaluated at $\omega = \omega_0$.

From (1) it can be obtained the velocity at which the energy in a pulse travels along a fiber (*group velocity*) and the *group-velocity dispersion* parameter, known as the GVD, which determines how much an optical pulse would broaden on propagation inside the fiber. They are defined as

$$v_g = \frac{1}{\left.\dfrac{\partial\beta(\omega)}{\partial\omega}\right|_{\omega=\omega_0}} = \frac{1}{\beta_1} \quad ; \quad GVD = \left.\frac{\partial^2\beta(\omega)}{\partial\omega^2}\right|_{\omega=\omega_0} \equiv \beta_2 \tag{2}$$

In some optical communications systems, the frequency spread $\Delta\omega$ is determined by the range of wavelengths $\Delta\lambda$ emitted by the optical source. It is customary to use $\Delta\lambda$ in place of $\Delta\omega$. By using $\omega = 2\pi c/\lambda$, 24) can be written as

$$D = -\frac{2\pi c}{\lambda^2}\beta_2 \tag{3}$$

where D is called the *dispersion parameter* and is expressed in units of ps/(km-nm).

In the fourth term of (1), the factor of β_3 is known as the *third-order dispersion*. This term is important around the wavelength at which β_2 is equal to zero (*zero-dispersion wavelength*). Higher-order dispersive effects are governed by the dispersion slope $S = dD/d\lambda$. The parameter S is also called a *differential-dispersion* parameter. By using (3), it can be written as

$$\beta_3 = \frac{\partial\beta_2}{\partial\omega} = -\frac{\lambda^2}{2\pi c}\frac{\partial\beta_2}{\partial\lambda} = -\frac{\lambda^2}{2\pi c}\frac{\partial}{\partial\lambda}\left[-\frac{\lambda^2}{2\pi c}D\right] \tag{4}$$

$$S = \frac{4\pi c}{\lambda^3}\beta_2 + (\frac{2\pi c}{\lambda^2})^2\beta_3 \quad \left[\frac{ps}{(nm)^2 \cdot km}\right] \tag{5}$$

The numerical value of the dispersion slope S plays an important role in the design of CWDM systems. Since $S > 0$ for most fibers, different channels have slightly different GVD values. This feature makes it difficult to compensate dispersion for all channels simultaneously.

3. Directly modulated laser

DFB lasers are the workhorse in WDM systems, both as cooled and uncooled devices. In order to provide cost-effective solutions for CWDM systems, directly modulated, uncooled distributed feedback (DFB) lasers are preferred (DFB-DML).

As regards cost, CWDM systems are usually single-span (unlike their DWDM counterparts), inasmuch as they do not use any type of in-line optical amplification. Since there is no optical amplification, the output power of the laser needs to be sufficient to sustain the system's loss budget. This involves the transmission fiber, multiplexer loss together with additional splice and connector losses.

A typical N-wavelength CWDM system, as standardized in ITU-T G.695, is shown in Fig. 1.

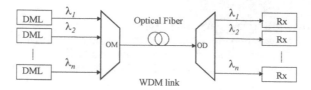

Fig. 1. Proposed optical transmission system.

The transmitter block diagram implemented in our studies consists of a bit random generator, which determines the sequence of bits, a_k, that will be sent to an electric pulse generator, in NRZ format, which injects a modulation current, $I(t)$ into the laser diode (DML), uncooled. A block diagram of the directly modulated transmitter to be considered is illustrated in Fig. 2.

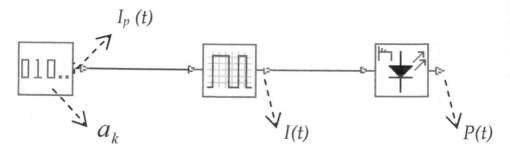

Fig. 2. Directly modulated laser scheme.

The injected laser current is given by the expression (Hinton & Stephens, 1993)

$$I(t) = I_b + \sum_{k=-\infty}^{\infty} a_k I_p(t - kT) \tag{6}$$

where I_b is the bias current, T is the period of the modulation pulse, a_k is the sequence of bits transmitted ($a_k = 1$ (0) if a binary one (zero) is transmitted during the k_{th} time), and $I_p(t)$ is the applied current pulse.

In a free-chirp source, the optical power output pulse of the laser, $P(t)$, is given by

$$P(t) = \eta_0 \cdot \frac{hv}{q} \cdot \sum_{k=-\infty}^{+\infty} a_k I_p(t - kT) \tag{7}$$

where η_0 is the differential quantum efficiency of the laser, hv the photon energy at the optical frequency v, and $I_p(t)$ the applied current pulse.

However, expression (7) is not applicable in case of directly modulated sources where the injected current that modulates the laser introduces a shift in the emission frequency (chirp frequency). As a consequence, the optical power output pulse is not a linear transformation of the applied current pulse.

3.1 Power and chirp waveform from the laser

The optical power and chirping response of the semiconductor laser to the current waveform $I(t)$ is determined by means of the large-signal rate equations, which describe the interrelationship between the photon density, carrier density, and optical phase within the laser cavity.

To calculate the optical power at the laser output depending on the value of the modulation current, we take into account the equations for the dynamics of the laser, which are given by the relationships between the density of photons $S(t)$, density of carriers $N(t)$ and the phase of the electrical field $\phi(t)$. In this way, the variation in carriers is related to the current waveform injected into the active layer $I(t)$ (Cartledge & Srinivasan, 1997; Cartledge et al. 1989; Corvini, 1987), in accordance with the following expression

$$\frac{dN(t)}{dt} = \frac{I(t)}{qV} - G[N(t) - N_t]S(t) - \frac{N(t)}{\tau_n} \tag{8}$$

and the variation of photons is given by:

$$\frac{dS(t)}{dt} = \Gamma G[N(t) - N_t]S(t) - \frac{S(t)}{\tau_p} + \frac{\beta \Gamma N(t)}{\tau_n} \quad ; \quad G = g_0 / [1 + \varepsilon S(t)] \tag{9}$$

where N_t is the carrier density at transparency, q the electron charge, V the volume of the active region, τ_n the electron lifetime, Γ the mode confinement factor in the cavity, β is the fraction of spontaneous emission coupled to the laser mode, τ_p the photon lifetime and G is the gain non-linear coefficient, ε being the compression factor of gain, g_0 the gain slope constant given by $g_0 = a_0 v_g$, where a_0 is the active layer gain coefficient and v_g the group velocity.

Thus, direct modulation of the laser current not only imprints the desired transmit data onto the emitted optical intensity, but also leads to a modulation of the carrier density within the laser cavity and in turn to a modulation of the refractive index of the active region. This in turn leads to the modulation of the frequency of the emitted light, a phenomenon that is known as laser chirp (the emission frequency must fulfil the phase condition all the time in order for the laser oscillation to occur). The chirp supposes that, even with singlemode lasers, the value of the instantaneous frequency of the emission is not the same during the optical pulse.

The frequency chirp Δv (deviation of the optical frequency at around the unmodulated frequency) is described by the coupled rate equations of the carriers and the optical field in the laser and their analytic solution yields an expression:

$$\Delta v = \frac{1}{2\pi} \frac{d\phi}{dt} \tag{10}$$

ϕ being the optical phase, whose variation with the time depends on the linewidth enhancement factor or the Henry factor α (Henry, 1982):

$$\frac{d\phi}{dt} = \frac{1}{2}\alpha \left[\Gamma v_g a_0 (N(t) - N_t) - \frac{1}{\tau_p} \right] \tag{11}$$

The linewidth enhancement factor α describes the relationship between how the real and imaginary refractive indices are affected by the carrier density. Factor α provides a measure of the coupling strength between the intensity and frequency modulation (FM) of the laser diode. It is defined as (Coldren & Corzine, 1995)

$$\alpha \equiv -\frac{dn/dN}{dg/dN} \tag{12}$$

Here, n is the real part of refractive index of the laser cavity. dg/dN is also known as the differential gain. From (10) and (11), the amount of frequency modulation is proportional to the linewidth enhancement factor which is typically between 4-10 but can be as low as 1 in some active materials.

The output power $P(t)$ is related to the photon density $S(t)$ through the expression

$$P(t) = \frac{V\eta h\nu}{2\Gamma \tau_p} \cdot S(t) \tag{13}$$

where h is the Planck constant and η is the differential quantum efficiency.

From the above equations the expression that relates the chirp with the laser output power can be obtained

$$\Delta\nu = \frac{\alpha}{4\pi}\left[\frac{1}{P(t)} \cdot \frac{dP(t)}{dt} + \kappa P(t) \right] \tag{14}$$

κ being the adiabatic chirp coefficient, which is directly related to the non-linear gain compression factor

$$\kappa = \frac{2\Gamma}{V\eta h\nu} \cdot \varepsilon \tag{15}$$

The first term in (14), is a structure-independent transient chirp and it produces variations in the pulse temporal width, while the second term is a structure-dependent adiabatic chirp, that produces a different frequency variation between the 1 and 0 bits (if the frequency of "1" is greater than the "0", it means a blue shift) causing a shift in time between the levels corresponding to "1"and "0" when the pulses go through a dispersive media, such as the optical fiber (Koch, 1988; Hakki, 1992).

DMLs can be classified according to their chirp behavior; Transient-chirp dominated DML when the first term of (14) is predominant and adiabatic-chirp dominated DML when the first term of (14) can be ignored compared to the second term of (14) (Hinton & Stephens, 1993). In following sections, we have called a DMLs with a strongly transient chirp dominated behavior DML-T and DML-A when its behavior is strongly adiabatic chirp dominated.

Fig. 3 shows the simulated power (a and c) and chirp (b and d) waveforms for two different DMLs. In this simulation, the modulating current $I(t)$ was made up of 2.5-Gb/s Gaussian current pulses.

In the case of an adiabatic chirp dominated laser (DML-A), 1 and 0 bits have a different frequency proportional to their optical power. The "0" frequency is smaller than the "1"' frequency [see Fig. 3 (b)]. While, in the case of an adiabatic chirp dominated laser (DML-T), the power waveform shows a large power overshoot on the 1's [see Fig. 3 (c)], and the chirp has a significant value during transition states. Both behaviors are consistent with (14)

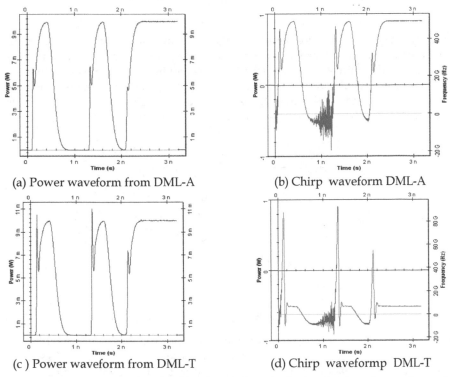

(a) Power waveform from DML-A   (b) Chirp waveform DML-A

(c) Power waveform from DML-T   (d) Chirp waveformp DML-T

Fig. 3. Simulated power (left-hand column) and chirp (right-hand column) waveforms for 2.5-Gb/s DML-A (a and b) and DML-T (c and d).

It is important to highlight that through (6-14) the frequency chirp for large-signal modulation can be determined directly from the shape of the modulated signal (e.g, square-wave, sine-wave, Gaussian, etc.). This simple expression is useful when discussing chirping phenomena because it clearly shows the interdependence of intensity and frequency modulation. In particular, the relative phases of frequency and intensity excursions are clearly displayed by (14), and these are of utmost importance in understanding the behavior of chirped signal transmission in dispersive media.

4. Transmission characteristics

When an optical pulse is transmitted through a dispersive media, such as an optical fiber, the intensity and shape of the optical signal at the output of the optical fiber, due to the waveform $I(t)$, are related with the dispersive characteristics of the optical fiber. This occurs because the spectral components that constitute the pulse are attenuated and/or phase shifted by different amounts. The effect of dispersion is more dramatic for ultrashort pulses since they have greater spectral widths.

The basic propagation equation that governs pulse evolution inside a single-mode fiber, considering expanding β to the third order of (1), is given by (Agrawal, 2010)

$$\frac{\partial A}{\partial z} + \beta_1 \frac{\partial A}{\partial t} + j\frac{\beta_2}{2}\frac{\partial^2 A}{\partial t^2} - \frac{\beta_3}{6}\frac{\partial^3 A}{\partial t^3} = 0 \tag{16}$$

Upon propagation in a linear lossless dispersive medium, a monochromatic plane wave of frequency ν traveling a distance z in the z direction undergoes a phase shift and the pulse will be broadened. Figure 4 shows the propagation of an initially unchirped Gaussian pulse through a fiber with anomalous dispersion (β_2 = GVD < 0). The pulse remains Gaussian, but its width expands and it becomes chirped with a decreasing chirp parameter (down-chirped). In this case the pulse has a decreasing instantaneous frequency.

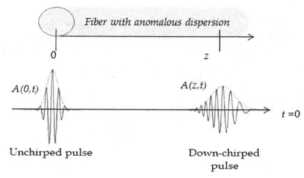

Fig. 4. Transmission of an optical pulse through a dispersive medium.

Thus, a fiber with a GVD parameter converts an unchirped Gaussian pulse of width τ_1 into a chirped Gaussian pulse of width τ_2 and chirp parameter a. Therefore, it follows that the process of pulse propagation through an optical fiber is equivalent to a combination of a time delay (second term of (16), group velocity) and a phase filter with chirp parameter b (third and fourth terms of (16) proportional to the distance z. By ignoring the third-order dispersion, the chirp parameter can be written in the form (Saleh & Teich, 2007)

$$b = 2\beta_2 z = \frac{GVD}{\pi}z \tag{17}$$

A fiber with $\beta_2 > 0$ ($b > 0$) is said to have normal dispersion (negative dispersion coefficient, D) and it functions as an up-chirping filter which increase its instantaneous frequency. If $\beta_2 < 0$ ($b < 0$) is said to have anomalous dispersion (positive dispersion coefficient, D) and it corresponds to a down-chirping filter which decreasing its instantaneous frequency.

In normal dispersion, a blue wavelength is slower than red, so the leading edge of the pulse is red-shifted and the trailing edge is blue-shifted. The opposite occurs for anomalous dispersion.

Based on (16) and considering both the case in which the carrier wavelength is far away from the zero-dispersion wavelength (contribution of the β_3 term is negligible) and the non-linear effects in the fiber to be negligible, the fiber can be seen as a phase filter with a chirp parameter b according to (17), thus:

- For an optical fiber with normal dispersion ($b > 0$) the filter is up-chirping.
- For an optical fiber with anomalous dispersion ($b < 0$) the filter is down-chirping.

A standard single mode fiber (SSMF), optimized for the transmission in the *O-band* (~1310 nm), is a typical fiber with anomalous dispersion whose transmission characteristics are described in *Recommendation ITU-T G.652*. The negative *non-zero dispersion shifted fiber* (N-ZDSF), whose chromatic dispersion coefficient is negative in the *C-band* (and above), is an example of a normal dispersion fiber (*Recommendation ITU-T G.655*).

5. The relationship between laser chirp and dispersion

The pulse shape at a distance z inside the fiber is a function of different factors: the initial pulse shape, the chirp associated with this pulse from the DML and the dispersive characteristics of the optical fiber.

As already mentioned, for wavelengths far away from the zero-dispersion wavelength and considering non-linear effects in the fiber as negligible, the fiber transmission can be regarded as a phase filter with a chirp parameter b in accordance with (17) and with a transfer function given by (Saleh & Teich, 2007)

$$H_e(f) = \exp\left[-jb\left(\pi f\right)^2\right] \tag{18}$$

In the following sections the influence of the interplay between chirp and dispersion on a propagated pulse is examined by regarding the process of propagation as a phase filter with a transfer function governed by (19) and considering that, in accordance with (14) and (16), the optical output power from a DML is blue shifted during the pulse rise time (increasing frequency) and red shifted during the pulse fall time (decreasing frequency), that is, the pulse is a down-chirped pulse.

Firstly, we are going to consider the propagation of a chirped-Gaussian. Although the Gaussian pulse has an ideal shape that is not encountered exactly in practice, it is a useful approximation that lends itself to analytical studies. In section 5.2 a general formula is derived that can be used to evaluate the waveform of a chirped pulse of arbitrary shape.

5.1 Propagation of chirped-Gaussian pulses in an optical fiber

A general Gaussian pulse has a complex envelope $A(z,t)$

$$A(z,t) = A_0 \exp\left[-\frac{1}{2}(1 - ia_1)\frac{t^2}{T_0^2}\right] \tag{19}$$

where A_0 is the peak amplitude, T_0 represents the half-width of the pulse at the $1/e$ power point. The parameter a_1 in (19) takes into account the frequency chirp of the DML at $z = 0$ (ignoring the adiabatic chirp, $\kappa = 0$).

When a chirped Gaussian pulse, defined by (19), is transmitted through a phase filter, with a transfer function given by (18), the outcome is also a chirped Gaussian pulse. Thus, the shape waveform of the filtered pulse, after a distance z, is given by the equation governing the evolution of the complex amplitude of the field envelope along the fiber, $A(z, t)$

$$A(z,t) = \frac{A_0}{\sqrt{1-ja_2} \cdot \frac{T_1}{T_0}} \exp\left[-\frac{(1-ja_2)}{2T_1^2} t^2\right] \tag{20}$$

where

$$\frac{2T_1^2}{1-ja_2} = \frac{2T_0^2}{1-ja_1} + jb \tag{21}$$

T_1 is the half-width similar to T_0.

Equating the real and imaging parts of (21) leads the width T_1 and the chirp parameter a_2 associated with the pulse after a distance z

$$b_f(z) = \frac{T_1}{T_0} = \left[\left(1 + \frac{a_1 b}{2T_0^2}\right)^2 + \left(\frac{b}{2T_0^2}\right)^2\right]^{1/2} = \left[\left(1 + \frac{a_1 \beta_2 z}{T_0^2}\right)^2 + \left(\frac{\beta_2 z}{T_0^2}\right)^2\right]^{1/2} \tag{22}$$

$$a_2 = a_1 + \left(1 + a_1^2\right)\frac{b}{2T_0^2} \tag{23}$$

Therefore, when a Gaussian pulse with an initial chirp a_1, travels a distance z in a dispersive media, the outcome is also a chirped Gaussian pulse with altered parameters. The chirp parameter will acquire a new value given by a_2 and the pulse will be broadened (or compressed) by a factor b_f.

In summary, as expected, the shape of the signal at the output of the optical fiber is affected by the sign of both β_2 (fiber parameter) and a (transmitter parameter). Compression can occur if an up-chirped pulse travels through a down-chirping (anomalous) medium, or if a down-chirped pulse travels through an up-chirping (normal) medium.

Taking into account the relationship between β_2 and D given by (5), Fig. 5 shows the pulse broadening factor b_f as a function of the accumulated dispersion, DL for an initial chirped Gaussian pulse from a DML with $T_0 = 30$ ps (equivalent to the bit period for a 10-Gb/s signal), dynamic chirp only, and a wavelength $\lambda = 1550$ nm and different chirp values (0, ±2 y ±4). In this case, the chirp parameter a_1 in (23) is the linewidth enhancement factor α given by (12). In order to be congruent with the notation used here, a_1 is taken as -α.

When comparing the result for the chirped pulses with $\alpha = 2$ or 4 (DML) with the result for the unchirped pulse ($\alpha = 0$), it is observed that for $D > 0$ (SMF fiber), the presence of chirp leads to significantly increased dispersion-induced pulse broadening, which increases in line with the increased chirp. However, for up to a certain amount of negative dispersion (N-NZDSF fiber), the chirp is actually beneficial and leads to pulse compression so that one can enhance the transmission system behavior.

From Fig. 5, it is observed that there is an optimum value for the accumulated dispersion that leads to minimum-width pulse. To determine the value b_{min} at which the pulse has its minimum width T_1^{min}, the derivative of T_1 in (22) with respect to b is equated to zero. The result is (Saleh & Teich, 2007).

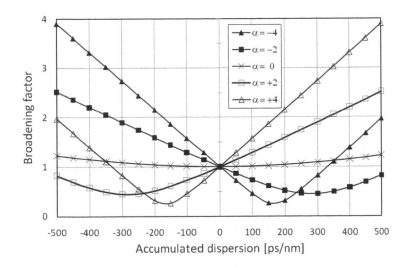

Fig. 5. Calculated pulse broadening factor b_f as a function of the accumulated dispersion: DL.

$$T_1^{min} = \frac{T_0}{\sqrt{1+a_1^2}}$$

$$b_{min} = -a_1\left(T_1^{min}\right)^2 = -2\frac{a_1}{1+a_1^2}T_0^2 \qquad (24)$$

From (245) we can calculate the fiber length at which the width T_1^{min} is reached

$$z_{min} = -\frac{a_1}{\beta_2\left(1+a_1^2\right)}T_0^2 \qquad (25)$$

Based on (25) z_{min} only exists for $a_1\beta_2 < 0$.
As an example, Fig. 6 (a) shows the simulated waveform of a positive chirped Gaussian pulse together with the chirp pulse when it is propagated through a negative dispersion fiber for three fiber lengths. Fig. 6 (b) represents the chirped Gaussian pulse and the phase throughout the pulse ($L_1 < L_2 < L_3$). The minimum width is reached at a length L_2 in accordance with (24). It is important to highlight that the minimum width T_1^{min} is obtained when the chirp value is zero (total compensation between DML-chirp and dispersion) and the accumulated phase is constant. Based on this, we will demonstrate the enhancement performance of an optical system by researching a modulation current waveform so that a constant accumulated phase, along the link, is achieved.

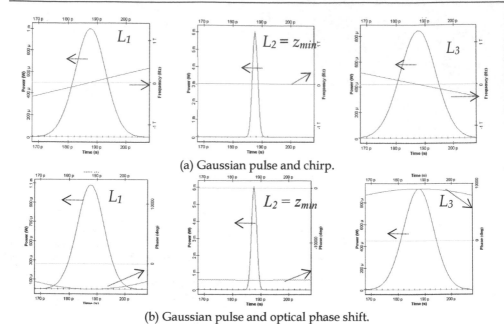

(a) Gaussian pulse and chirp.

(b) Gaussian pulse and optical phase shift.

Fig. 6. Simulated waveform of a positive chirped Gaussian pulse and chirp (a) when the pulse is propagated through a negative dispersion fiber for three fiber lengths. Chirped Gaussian pulse and phase throughout the pulse (b).

5.2 Chirped pulses of arbitrary shape

Equation (20) only describes the propagation of a Gaussian pulse and includes dispersive effects only up to the second order (β_2). In practice, the pulses cannot be considered Gaussian, mainly for two reasons; as is demonstrated in Section 3.1. The pulse waveform of a DML exhibits an overshoot and ringing in its output power whilst, on the other hand, if a higher-order dispersion governed by β_3 is included in (20), the pulse no longer remains Gaussian on propagation because it develops a tail with an oscillatory structure (Agrawal, 2010).

For pulses of an arbitrary shape, a measurement of the pulse width is provided by the *root mean square* (RMS) width of the pulse defined as:

$$\sigma_p = \left(\left\langle t^2\right\rangle - \left\langle t\right\rangle^2\right)^{1/2} \quad ; \quad \left\langle t^m\right\rangle = \frac{\int\limits_{-\infty}^{\infty} t^m \left|A(z,t)\right|^2 dt}{\int\limits_{-\infty}^{\infty} \left|A(z,t)\right|^2 dt} \tag{26}$$

Then, the broadening factor is given by

$$\frac{\sigma^2}{\sigma_0^2} = \left(1 + \frac{a\beta_2 z}{2\sigma_0^2}\right)^2 + \left(\frac{\beta_2 z}{2\sigma_0^2}\right)^2 + \left(\frac{\beta_3 z\left(1 + a^2\right)}{4\sqrt{2}\sigma_0^3}\right)^2 \tag{27}$$

where σ_0 is the input pulse ($\sigma_0 = T_0/\sqrt{2}$). The last term in (27) is the third-order dispersion contribution to the pulse broadening.

If the complex envelope of the optical field at the laser output is given by (Agrawal, 2010)

$$A(0,t) = \sqrt{P(t)}\exp[j\phi(t)] \qquad (28)$$

where $P(t)$ is the laser output power and $\phi(t)$ is obtained by the integration of (11).

The output signal complex envelope is obtained by the convolution of the input envelope (28) with the fiber response $h_e(z,t)$ [Fourier transform of (18)]. Thus, the output power at a z point is calculated as

$$P_z(t) = |A(z,t) * h_e(z,t)|^2 \qquad (29)$$

Starting from (29) the shape of a chirped pulse can be determined when it is propagated through fibers with both positive and negative dispersion values.

In general, (29) cannot be easy to solve in an analytic form. In this case, computer simulations are useful to predict the exact influence of the interplay of chirp and dispersion on transmission performance.

The purpose of the following section is to present simulation results pertaining to the transmission of chirped optical signals in a dispersive fiber. In the following, it will be shown that transient-chirp dominated lasers perform better over negative and positive dispersion fibers than adiabatic-chirp dominated lasers. It will be also shown that the transient component of the chirp improves the transmission performance significantly over positive dispersion fibers if the modulation current waveform is chosen appropriately.

6. System model

Good modulation performance at high bit-rates, to avoid excessive back-to-back penalties, is one of the main requirements for CWDM transmitters. Since the a parameter (including the α factor and κ coefficient) is tunable by means of the modulating current $I(t)$. The optimum operating conditions in terms of the chirp/dispersion interactions can be set for fibers having different amounts and signs of dispersion. It is the aim of this section to discuss and compare how the shape of the modulated signal (e.g. square-wave, sine-wave, Gaussian, etc.) can improve the system performance when using both positive and negative dispersion fibers.

As already mentioned, numerical simulation will be required to model the exact influence of the interplay between chirp and dispersion on transmission performance. The decision on the choice of the transmission fiber characteristics (i.e., absolute dispersion value and its sign) and the DML characteristics (adiabatic or transient) for Metro applications should first be determined through simulations. The parameters of the components involved in the simulation should be sufficiently accurate and representative of the majority of commercially available components, so that useful conclusions on the design and performance of the real system can be obtained. Our simulations are based on commercial software for optical transmission systems and the component parameters used in it have been extracted from the datasheet of trademarks.

Fig. 1 shows the basic CWDM link implemented in our simulations. In this work we are specifically interested in how an optical system is impacted by the pulse shape of modulation current. But our final purpose is to optimize a CWDM system for use in a metropolitan area optical network. Due to the chirping frequency, when a directly modulated laser is used, the system performance is very sensitive to the presence of spectrally selective components (such as multiplexers, demultiplexers and filters) in the link. As has already been shown analytically in previous sections, the shape of the pulse and its accumulated phase change as a result of their transmission through these components. For this reason, we have simulated the WDM link as shown in Fig. 1, including the optical multiplexer (OM) and demultiplexer (OD) components. Nevertheless, for the sake of clarity in analyzing the results, only the data related to one channel (i.e. the channel allocated at 1551 nm) are presented here. A summary of the most representative parameters used in our simulation is detailed below.

6.1 Optical fibers

As already mentioned, the chromatic dispersion coefficient, both in absolute terms and in sign, is one of the most influential fiber parameters in transmission performance when a DML is used as a transmitter. For this reason, in this work, we have simulated two fibers with opposite signs in their dispersion coefficient; the already laid and widely deployed, single-mode ITU-T G.652 fiber (SMF) and the ITUT-T G.655 fiber (N-NZDSF). It is well known that the SMF fiber dispersion coefficient is positive in the telecommunication band spectrum, from the O-band to the L-band, and the dispersion coefficient sign of the N-NZDSF fiber is negative around C-band.

In this work, spectral attenuation coefficient, dispersion slope, effective area and the non-linear index of refraction are compliant with SMF-28e and MetroCor Corning® commercial fibers. These fibers are in compliance with ITU-T G.652D and ITUT-T G.655, respectively.

Table I summarizes the values of the extracted parameters for the two fibers at 1551 nm wavelength.

	SMF	N-NZDSF
Attenuation Coefficient (dB/km)	0.20	0.20
Dispersion [ps/(nm·km)]	18	-5.6
Zero dispersion wavelength (nm)	1313	1605
Effective Group Index of Refraction	1.4682	1.469

Table I. Parameters for the two fibers at 1551 nm wavelength.

6.2 Transmitter

Distortion in the dispersion-induced waveform is the greatest hindrance that the designer of metro area networks has to contemplate. This distortion in the dispersion-induced waveform can have serious negative effects on the signal transmission, even at very short distances depending on the optical transmitter chosen and the characteristics of its frequency chirp.

The transmitter block diagram implemented in our simulations is shown in Fig. 2. The pulse string was encrypted with a $2^{15}-1$ pseudo-random bit sequence, OC-48 system, at 2.5 -Gb/s.

For our purpose, the leading and tailing edges of the applied current pulse $I_p(t)$ have been matched to different forms; exponential, sine and Gaussian (see Fig. 7) .

Exponential: Gaussian: Sine:

$$I_p(t)=\begin{cases}1-e^{-(t/c_r)} \to 0 \le t \le t_1\\1 \to t_1 \le t \le t_2\\e^{-(t/c_f)} \to t_2 \le T_b\end{cases} \quad I_p(t)=\begin{cases}e^{-(t/c_r)2}-1 \to 0 \le t \le t_1\\1 \to t_1 \le t \le t_2\\e^{-(t/c_f)^2} \to t_2 \le T_b\end{cases} \quad I_p(t)=\begin{cases}sin(\pi t/c_r) \to 0 \le t \le t_1\\1 \to t_1 \le t \le t_2\\sin(\pi t/c_f) \to t_2 \le T_b\end{cases} \quad (30)$$

where, c_r and c_f are the rise and fall time coefficients, respectively. t_1 and t_2, together with c_r and c_f, are numerically determined to generate pulses with the exact values of the rise time and fall time parameters, and T_b is the bit period. In all simulated cases, fall and rise time are fixed at the half bit period.

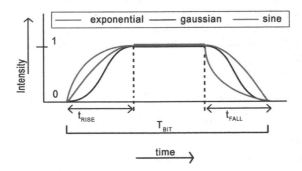

Fig. 7. Exponential, Gaussian and sine pulse shape for the laser modulation current.

To simulate the DMLs, the Laser Rate Equations introduced in (Cartledge & Srinivasan, 1997; Tomkos et al., 2000) has been used as model.

	DML-A	DML-T
Factor α	2.2	5.6
Volume active layer (cm ^{3}), V	$5.8\ 10^{-12}$	$1.1095\ 10^{-10}$
Gain compression coefficient (cm³), ε	$5\ 10^{-18}$	
Quantum efficiency, η	0.19	
adiabatic coefficient, κ (Hz/W)	$28.7\ 10^{12}$	$1.5\ 10^{12}$

Table II. Simulation parameters for directly modulated lasers DML-A and DML-T.

In the simulation, we have chosen the values of commercial lasers to generate two types of chirp behavior: DML-A with a dominant adiabatic chirp and DML-T with a dominant transient chirp. For this, we work with two typical values for the Henry coefficient, (α) (Yu & Guido, 2004; Villafranca et al, 2008; del Rio & Horche, 2008) and the adiabatic coefficient (κ) have been determined from the parameters ε, η, and Γ, in accordance with (17). These values are detailed in Table II.

6.3 Other components
The optical receiver was modeled as a *pin* photodiode with a *Responsivity* of 1 A/W and with a thermal noise of $1.84375*10^{-22}$ W/Hz, followed by a low-pass Bessel electric filter to eliminate high-frequency noise.

The WDM multiplexer and demultiplexer components are configured as optical Bessel filters (2nd order) with a bandwidth of 10 GHz. They are fixed in all simulations.

6.4 Evaluation of system performance

Four different systems have been analyzed in the computer simulations, all based on the optical link shown in Fig. 1; using a DML-A in conjunction with SMF and N-NZDSF fibers, and the same scheme by substituting the DML-A by the DML-T. The performance of the systems is evaluated and the results are compared.

The transmission system performance is often characterized by the *Bit Error Rate* (BER), which is required to be less than 10^{-15} for systems with a bit rate greater than 2.5-Gb/s. Experimental characterization of these systems is difficult since the direct measurement of the BER involves considerable time and cost at such a low BER value. Another way of estimating the BER is by calculating the Q-factor of the system, which can be modeled more easily than the BER. The BER, with the optimum setting for the decision threshold, depends on the Q-factor as (Alexander, 1997)

$$ BER = \frac{1}{2} erfc\left(\frac{Q}{\sqrt{2}}\right) \approx \frac{1}{\sqrt{2\pi}} \frac{e^{-\frac{Q^2}{2}}}{Q}; \qquad Q = \frac{|I_1 - I_0|}{\sigma_1 + \sigma_0} \qquad (31) $$

where, I_i and σ_i are average and variance values, respectively, of the 1 and 0 bits.

On the other hand, the Q-factor can be determined from *eye diagram*. The eye diagram provides a visual way of monitoring the system performance: The "*eye opening*" is affected by the dispersive effects accumulated along inside the link and a closing of the eye is an indication that the system is not performing properly. If the effect of laser chirp is small, the eye closure Δ can be approximated by (Keiser, 2010)

$$ \Delta = \left(\frac{4}{3}\pi^2 - 8\right) t_{chirp} DLB^2 \delta\lambda \left[1 + \frac{2}{3}\left(DL\delta\lambda - t_{chirp}\right)\right] \qquad (32) $$

where t_{chirp} is the duration of the chirp, B is the bit rate, D the fiber chromatic dispersion, L is the fiber length and $\delta\lambda$ is the maximum wavelength excursion induced by the chirp.

From (32), one approach to minimize the chirp effect is to work at the fiber zero dispersion wavelength. However, in WDM systems each channel has its own dispersion coefficient and it is not possible to implement this solution. Another way of minimizing the chirp effects, whatever the working wavelength, has to be found.

In the following section, the performance of the simulated systems is evaluated in terms of the Q-factor.

7. Simulations and results

For the simulation of the proposed system (Fig. 1), 100 km of optical fiber was used and, as already mentioned, only the results for the channel allocated at a wavelength of 1551 nm are presented. In all simulated cases the laser was polarized with a bias current slightly below the threshold; which means that no impact on the extinction ratio is considered.

According to (14), the DML chirp characteristics are dependent on their output optical power. In our previous works (Horche & del Rio, 2008, del Rio et al., 2011), we demonstrated the existence of an optimum optical power that leads to a maximum *Q-factor* (Q_{max}). This is why the study includes an analysis of the systems at different output optical powers of the DML. In commercial uncooled DFB-DMLs the average modulated output power can reach 6 dBm or more at room temperature. In our simulations, a peak power for the "1" bit between 0.1 mW to 10 mW (average ~ 6 dBm) has been considered.

For each of these cases the system performance (Q-factor) has been analyzed taking into account the three different modulations current types given by (30).

7.1 Transmitter with dominant adiabatic chirp, DML-A

In a DML, the output optical power waveform is a function of the modulation current, as a result of using different currents, different optical pulse shapes are obtained. Fig. 8(a) shows the output optical power pulse shape of a DML-A when the modulation current takes exponential, Gaussian and sine forms, respectively.

The three pulses have the typical characteristic shape of the pulses generated by an adiabatic chirp-dominated laser. As can be seen in Fig. 8(a), for the current Gaussian modulation shape, the pulse leading edge is delayed in relation to the exponential and sine forms. This is the result of a higher *turn-on delay* (turn-on delay is the time required for the carrier density to build up to the threshold value before light is emitted). As seen in the following sections, this is one of the reasons why the system with a Gaussian current pulse presents the worst performance.

In the case of an adiabatic chirp-dominated laser, where the transient chirp has been completely "masked" by the adiabatic term, there is a shift between the frequency of the 1 and the 0 bits, the frequency of the "1" being greater than the frequency of the "0" (blue shift). Therefore, the result of the dispersion with the specific chirp characteristics is a high-intensity spike at the front of the pulses and a trailing tail-end for transmission over a positive dispersion fiber (SMF), see Fig. 8 (b). Exactly the opposite effects take place for transmission over a negative dispersion fiber (N-NZDSF), see Fig. 8 (c) [Hakki, 1992]. In the case of DML-A the main role in the transmission performance is played by the absolute value of the dispersion coefficient (rather than its sign) [Tomkos et al., 2001]. Both figures, 8(b) and 8(c), take into account the simulation of a point-to-point optical link without any selective component in wavelength, that is, only the fiber introduces changes into the chirp of the transmitted pulse.

In Fig. 9 the maximum Q-factor, minimum BER, of the system in Fig. 1 is shown. Variations in the DML optical power from 0.1 to 10 mW for the three types of modulation current are analyzed for an SMF, Fig. 9(a), fiber and an N-NZDSF fiber, Fig. 9 (b). The graphs show that, regardless of the type of fiber used, the system has a maximum performance for a given value of power ($P_{optimun}$) which agrees with the results obtained in previous works (Horche & del Rio, 2008; Suzuki, 1993). For low levels of channel power (P_{ch}), below the optimum power, the Q-factor increases with $P_{ch}(t)$, mainly because of a larger amount of power reaches the detector and the performance of the system is improved. For $P_{optimun}$ the interplay of the chirp with the dispersion leads to the best compensation between them. On the other hand, for P_{ch} higher than $P_{optimun}$, the chirp increases with the power which gives rise to a greater frequency shift between the "1" and "0" levels and linewidth broadening producing a greater optical pulse deformation.

Looking at Fig. 9(a) and (b) we can see that the system has a similar behavior with the sine and exponential models and they are far from the Gaussian model. These variations are a result of the similarities/differences that exist in the shape of the pulses at the output of the laser when the different models are considered [see Fig. 8(a)].

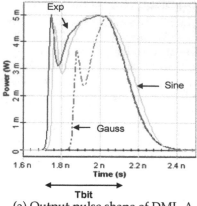

(a) Output pulse shape of DML-A.

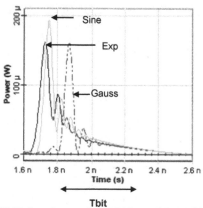

(b) Pulse shape after 100 km of SMF fiber

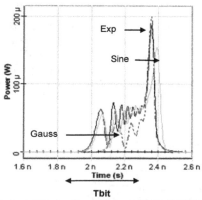

(c) Pulse shape after 100 km of N-NZDSF fiber

Fig. 8. Pulse shape for a DML-A with three types of modulation current: exponential, (thick dotted black line), Gaussian (thin dotted red line) and sine (continuous green line).

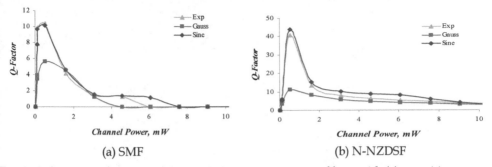

(a) SMF

(b) N-NZDSF

Fig. 9. Q-factor as a function of the DML-A output power on fibers with (a) a positive dispersion coefficient and (b) a negative dispersion coefficient.

Comparing the results from the SMF and N-NZDSF, Fig. 9(a) and (b) respectively, the N-NZDSF fiber outperforms the SMF fiber in all three cases of sine, exponential and Gaussian because of the reduced absolute dispersion value.

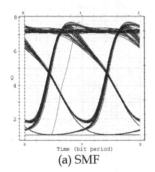
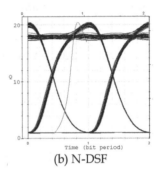

(a) SMF

(b) N-DSF

Fig. 10. Eye diagram for a DML-A laser with an optical output of 1 mW after 100 km of SMF (a) and N-NZDSF (b). A sine pulse is considered as the modulation current.

In Fig. 10, the eye diagrams are shown for the case of an adiabatic chirp dominated transmitter after transmission over 100 km of SMF (a) and N-NZDSF (b), respectively. In both case P_{ch} was 1 mW. The different dispersion sign will only affect the asymmetry of the eye diagram as is obvious from the results of Fig. 10.

From the simulations carried out with a DML-A transmitter, it can be concluded that with this type of laser the major factor is the absolute value of the fiber dispersion, and the waveform of the current modulation is not significant

7.2 Transmitter with a dominant transient chirp (DML-T)

Transient chirp-dominated laser diodes exhibit significantly more overshoot and ringing in output power and frequency deviations (see the power and chirp waveforms in Fig. 11). In case of a DML-T, the leading edge of the pulse is blue shifted relative to the main portion of the pulse, while the tailing edge is red shifted. The blue-shift chirped portion advances relative to the main portion of the pulses in the case of positive dispersion fibers. This effect produces a pulse spreading and consequently, an inter-symbol interference can occur.

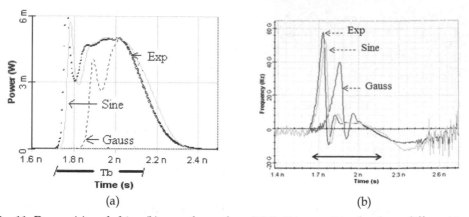

(a)   (b)

Fig. 11. Power (a) and chirp (b) waveforms for a DML-T transmitter for three different types of current modulation: exponential, Gaussian and sine.

Fig. 12 shows the Q-factor as a function of power for the three models studied. Fig. 12(a) corresponds to the propagation over an SMF fiber and Fig. 12 (b) is dealing with N-NZDSF. Clearly, from the results summarized in Fig. 12 some conclusions can be extracted:

- Again, the minimum quality standard is achieved for the Gaussian model
- In both SMF and N-NZDSF the existence of an optimum optical power for a maximum Q-factor is demonstrated. However, with the DML-T transmitter, the dispersion characteristics of the fiber used in the link determine the modulation current pulse shape for which the transmission performance is improved.

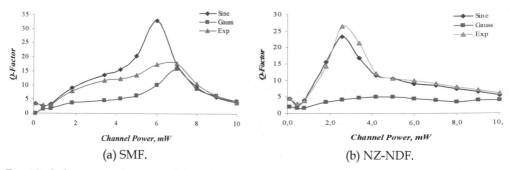

(a) SMF.   (b) NZ-NDF.

Fig. 12. Q-factor as a function of the DML-T output power on fiber with (a) a positive dispersion coefficient and (b) a negative dispersion coefficient.

Therefore, by using a negative dispersion fiber (Fig. 12(b)), the exponential current waveform is the best choice together with an output power of 2.6 mW. In this case, the reduction in quality that the sine and Gaussian pulse models have with respect to the exponential is approximately 11% and 84%, respectively.

When using an SMF fiber with a positive dispersion coefficient (see Fig. 12(a)), the modulation current with a sine shape has a better performance, particularly where the power of the laser is about 6 mW.

This improvement in performance observed with sine modulation current, can be understood in terms of the interaction that is produced between the DML-T chirp and pulse shape with the fiber dispersion; as is shown later, the accumulated phase in the received optical bit patterns is practically constant all of the time. When the chirped pulse travels along a fiber, it interacts with the fiber dispersion characteristics, causing a phase shift in the pulse, which is a function of the sign and the absolute value of the dispersion coefficient.

Fig. 13 shows the eye diagram after 100 km of SMF (a) and N-NZDSF (b) using a DML-T transmitter and when the modulation current has a sine shape. The eye corresponding to a transmission over an N-NZDSF fiber is more distorted than that corresponding to transmission over an SMF fiber. This is surprising because of the larger absolute value of the dispersion of SMF fiber compared with N-NZDSF fiber.

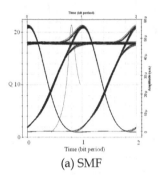
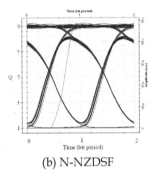

(a) SMF   (b) N-NZDSF

Fig. 13. Eye diagram for a DML-T laser with an optical output of 6 mW after 100 km of fiber. A sine pulse is considered as modulation current.

Through previous simulations, it has been demonstrated that directly modulated systems using SMF fibers can reach a similar (or better) Q-factor than those using N-NZDSF fibers, through the use of a correct modulation current. It means that through a suitable choice of emitted optical power in the DML together with the shape of the modulation current, the tolerance of the system at chirp can be optimized and its performance improved. This is an important conclusion since the SMF fiber is currently the most commonly installed, even when operating at 1550 nm.

8. Study of accumulated phase

All the features of the received waveforms analyzed in the previous section can be easily explained by considering the interaction of the laser chirp and pulse shape with the fiber dispersion. Then, the different Q-factors obtained with the sine, exponential and Gaussian models, when a DML-T is used together with an SMF fiber, can be justified simply by studying the accumulated phase in the bit sequence.

In the analysis of the accumulated phase, we must be very careful with the choice of the bit sequence length as the accumulated phase at reception depends on the number of transitions between the 0's and 1's in the bit pattern.

As discussed in section 3.1, the DML chirp depends on the emitted average power. Thus sending a short sequence of alternating 1 and 0 bits, the chirp fluctuates between a maximum value (when sending a "1") and a minimum (when sending a "0") and the accumulated phase will take a constant average value. Fig. 14(a) shows the accumulated phase for a bit sequence of 8 bits (01010101). However, if the bit sequence has several identical symbols, the results can be very different. For example, Fig. 14 (b) and (c) show the accumulated phase for the "11110000" and "00001111" sequences, respectively.

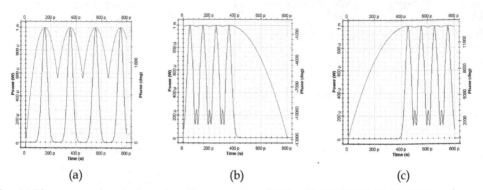

(a)   (b)   (c)

Fig. 14 The accumulated phase for a bit sequence of 8 bits (a) 01010101 (b) 1111000 and (c) 00001111.

Therefore, to obtain a result independent of the number of symbol transitions, we have to choose a large enough length sequence. On the other hand, this has the disadvantage of requiring excessively long processing times. In our simulation, we used a sequence of 512 bits, which maintains a good balance between accuracy of results and the calculation time.
Fig. 15 represents the accumulated phase variation over a sequence of 512 bits, for different DML-T output power for the sine model after 100 km of SMF fiber. As can be seen, there is a single value of power (6 mW) that produces a constant phase variation over time that would be the best choice for the transmission. For this power value the phase variation is constant in time with the accumulated chirp being practically zero.

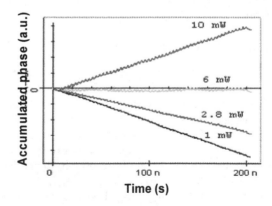

Fig. 15. Phase shifted (chirp) produced in a 512 bit sequence, for the sine model after 100 km of SMF fiber.

The accumulated phase is now represented for a sequence of 512 bits at the output of the DML-T laser with 6mW [Fig. 16(a)] and before the bit stream reaches the receiver [Fig. 16(b)], for the exponential, sine and Gaussian models. It can be seen that the bit stream goes through different accumulated phase variations when going along the SMF fiber and the filter components in the link (multiplexers, demultiplexers). The accumulated phase due to the sine pulse shape is practically constant, unlike that produced for exponential and Gaussian pulses.

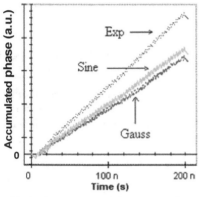

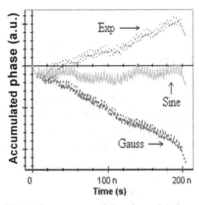

(a) At the output of the DML-T laser   (b) At the output of the demultiplexer.

Fig. 16. Accumulated phase shifted (chirp) for a DML-T transmitter with 6 mW of optical power, 512-bit sequence, both before and after 100 km of SMF fiber.

The high dispersion tolerance is mainly because of the phase-correlative modulation between the adjacent bits via a precise control of the frequency chirp in the DML modulation. Therefore, based on (1-8), we can design the parameters of the DMLs (including the drive current (DC) bias) to offer a suitable chirp response for the generation of the phase correlation.

9. Analysis of transmission performance through different SMF and N-NZDSF fiber lengths

In order to get a general idea of the behavior of the systems detailed in Fig. 1 with the link length, a set of simulations was carried out by varying the fiber length, L. As in the former cases, only the results for a channel centered at 1551 nm are shown. The simulated transmitter was a DML-T similar to that used in the former sections. The output power is P_{ch}= 6 mW in case of a SMF fiber [see Fig. 12 (a)] and P_{ch}= 2.6 mW when a NZ-NDF fiber is used [see Fig. 12 (b)]. These power values have been selected to obtain the Q_{max}.

Fig. 17 represents the Q-factor versus fiber length when an SMF fiber (a) and a NZ-NDF fiber (b) are used for sine, Gaussian and exponential modulation current waveforms, respectively. As expected, the Q_{max} is reached for a fiber length of 100 km and a sine modulation current in the case of an SMF fiber. This is the length at which the interaction between the accumulated dispersion in the fiber and the chirp generated in the DML is optimal. Nevertheless, for other fiber lengths, another choice can provide a better performance.

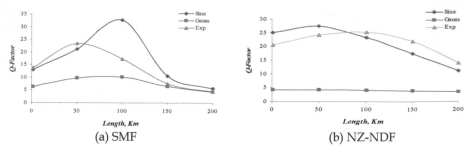

(a) SMF   (b) NZ-NDF

Fig. 17. Q-factor as a function of the link distance through fibers with (a) positive dispersion coefficient and (b) negative dispersion coefficient (Pch = 6 mW).

In the case of an N- NZDSF fiber, in Fig. 17 (b), with a P_{ch}= 2.6 mW, the Q_{max} is obtained for lengths different from 100 km, i.e. for a sine modulation current, L = 50 km [not exponential as in Fig. 12 (b)]. Therefore, we can conclude that the maximum of the Q factor depends on the modulation current shape, the length of the link and the optical power at the DML output.

10. Conclusion

In this chapter we have determined the dependence of the system performance of a directly-modulated optical system as regards the shape of the optical pulse transmitted through the fiber. It has been demonstrated that, if the laser has adiabatic dominant chirp behavior (DML-A), the absolute value of the fiber dispersion coefficient will be a determinant of the system performance but the influence of the current pulse shape is not significant.

We have also demonstrated that, if the laser has transient dominant chirp behavior, it is possible to improve the performance of an already installed system by modifying the shape of the electric current that modulates the laser. As the high dispersion tolerance is mainly because of the phase-correlative modulation between the adjacent bits, then, we can design the parameters of the DMLs, including the modulating current shape, to offer a suitable chirp response for the generation of the phase correlation. For SMF fibers (positive dispersion coefficient), this shape is sinusoidal and for NZ-NDF fibers (negative dispersion coefficient), the shape is exponential, when long-haul inter-office transport systems, corresponding to link distances of about 40 - 100 km, are considered.

In summary, through the appropriate combinations of DML transmitters (adiabatic or transient) and shape current, optical fiber systems can be achieved which are optimized in terms of dispersion and cost. With this method it is possible to improve each of the WDM system channels individually, offering a low-cost solution since it only involves changes in the transmitters and avoiding the replacement of the fiber.

11. Acknowledgements

The authors gratefully acknowledge the support of the MICINN (Spain) through project TEC2010-18540 (ROADtoPON).

12. References

Agrawal, P. (2010) Fiber-Optic Communication System. *John Wiley & Sons*. ISBN 978-0-470-50511-3.

Alexander S. B. (1997). Optical Communication Receiver Design. SPIE Press/ IEE.

Cartledge, J.C. & Burley, G.S., (1989). The effects of laser chirping on lightwave system performance, *J. Lightwave Technol.*, Vol. 7, No. 3, pp. 568–573, 1989. ISSN: 0733-8724

Cartledge, J.C. & Srinivasan, R.C. (1997). Extraction of DFB Laser Rate Equation Parameters for System Simulation Purposes. *J. of Lightwave Technol.*, Vol. 15, No. 5, May 1997, pp. 852-860. ISSN: 0733-8724.

Coldren L.A., & Corzine, S.W. (1995) *Diode Lasers and Photonic Integrated Circuits*, Wiley Series in Microwave and Optical Engineering.

Corvini, P. & Koch, T. (1987). Computer Simulation of High-Bit-Rate Optical Fiber Transmission Using Single-Frequency Lasers. *J. Lightwave Technol.*, Vol. 5, No. 11, pp. 1591 - 1595. November 1987. ISSN: 0733-8724.

del Río, C. & Horche, P.R. (2008). Directly modulated laser intrinsic parameters Optimization for WDM Systems. *International Conference on Advances in Electronics and Micro-electronics*, ENICS 2008.

del Río, C., Horche, P.R., & Martín, A. (2011), Interaction of semiconductor laser chirp with fiber dispersion: Impact on WDM directly modulated system performance". *Proc. of The Fourth International Conference on Advances in Circuits, Electronics and Micro-electronics*. CENICS 2011, 22-27 August, Nize, France.

Hakki, B.W. (1992). Evaluation of transmission characteristics of chirped DFB lasers in dispersive optical fiber. *J. of Lightwave Technology*, Vol. 10, No. 7, pp. 964 – 970, Jul 1992. ISSN: 0733-8724

Henry, C. H. "Theory of the linewidth of Semiconductor Lasers". *IEEE J. of Quantum Electronics*, Vol. QE-18, pp. 259 - 264, Feb 1982. ISSN: 0018-9197.

Hinton, K.; Stephens, T.; Modeling high-speed optical transmission systems. *IEEE J. Selected Areas in Communications*, Vol. 11, Issue 3, pp. 380 - 392.

Horche, P.R. & del Río, C.. "Enhanced Performance of WDM Systems using Directly Modulated Lasers on Positive Dispersion Fibers". *Optical Fiber Technology*. Vol. 14, No. 2, pp. 102-108, April 2008.

Keiser, G., (200). *Optical Fiber Communications*. McGraw-Hill. ISBN: 978-007-108808.

Koch, T.L.; Corvini, P.J., (1988). Semiconductor laser chirping-induced dispersive distortion in high-bit-rate optical fiber communications systems. *IEEE International Conference on Communications*, ICC '88. Vol. 2, pp. 584 - 587.

Nebeling, M. (2007). CWDM Transceivers, In: *Coarse Wavelength Division Multiplexing; Technologies and Applications*. Edited by Hans-Jörg Thiele & Marcus Nebeling, pp. (57-90), CRC Press Taylor & Francis Group, ISBN-10: 0-8493-3533-7, USA.

Saleh, B.E.A., Teich, M.C., Fundamentals of Photonics. (2007). Wiley Series in Pure an Applied Optics. ISBN: 978-0-471-35832-9, USA.

Suzuki, N., (1993). Simultaneous Compensation of Laser Chirp, Kerr Effect, and Dispersion in 10 Gp/s Long-Haul Transmission Systems. *J. of Lightwave Technology*, Vol. 11, No. 9, Sep 1993. ISSN: 0733-8724

Thiele, H-J & Nebeling, M (2007) *Coarse Wavelength Division Multiplexing; Technologies and Applications*. CRC Press Taylor & Francis Group, ISBN-10: 0-8493-3533-7, USA.

Tomkos I.; Roudas, I.; Boskovic, A.; Antoniades, N.; Hesse, R. & Vodhanel, R. (2000). Measurements of Laser Rate Equation Parameters for Simulating the Performance of Directly Modulated 2.5 Gb/s Metro Area Transmission Systems and Networks. *IEEE Lasers and Electro-Optics Society, LEOS 2000,* Vol 2, pp. 692 – 693, 2000.

Tomkos, I., Chowdhury, D., Conradi, J., Culverhouse, D., Ennser, K., Giroux, C., Hallock, B., Kennedy, T., Kruse, A., Kumar, S., Lascar, N., Roudas, I., Sharma, M., Vodhanel, R. S., & Wang, C.-C. (2001). Demonstration of Negative Dispersion Fibers for DWDM Metropolitan Area Networks, *IEEE J. on selected Topis in Quantum Electronics.* Vol. 7, No. 3, May/Jun 2001, pp. 439-460.

Villafranca, A., Lasobras, Escorihuela, J. R., Alonso, R. & Garcés, I. "Time-resolved Chirp Measurements using complex Spectrum analysis based on stimulated Brillouin Scattering" *Proceedings of the OFC,* Paper OWD4, San Diego (USA), Feb 2008.

Yu, Y., & Giuliani, G. (2004). Measurement of the Linewidth Enhancement Factor of Semiconductor Lasers Based on the Optical Feedback Self-Mixing Effect. *IEEE Photonics Technology Letters,* Vol. 16, No. 4, pp. 990 - 992. April 2004. ISSN:1041-1135

Linke, R.A. (1985). Modulation Induced Transient Chirping in Single Frequency Lasers. *IEEE J. Quantum Electronics,* Vol. QE-21, No. 6, (June 1985), pp. 593-597, ISBN

Morgado, J.A.P & Cartaxo, A.V.T. (2001). Dispersion supported transmission technique: comparation of performance in anomalous and normal propagation regimes. *IEE Proc-Optoeelctron.,* Vol. 148, No. 2, (April 2001), pp. 107-116, ISBN

Part 2

Sensors

5

Fiber Fuse Propagation Behavior

Shin-ichi Todoroki
National Institute for Materials Science
Japan

1. Introduction

A fiber fuse is the continuous self-destruction of optical fiber induced and fed by propagating light. It is triggered by the local heating of a waveguide structure through which a high power beam is being delivered. A typical example is seen along a single mode silica glass optical fiber delivering a few watts of light (see Fig. 1). Once heat-induced high density plasma (or an optical discharge) is captured in the core region, it travels along the fiber toward the light source, consuming the light energy and leaving a hollow damage train. Thus, this phenomenon imposes an inevitable limit on the light power per fiber for current optical communication systems, namely, the service be unable to meet the growing demands of communication traffic.

The first oral presentation describing the fiber fuse phenomenon was given by Kashyap (1988), followed by the first publication by Kashyap & Blow (1988). These appeared just after certain

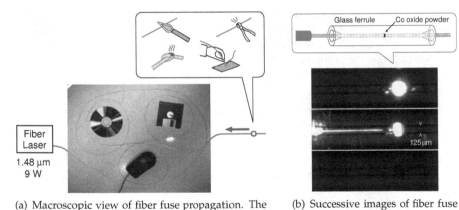

(a) Macroscopic view of fiber fuse propagation. The speed is about 1.2 m/s.

(b) Successive images of fiber fuse ignition (30 frames per second, Todoroki (2005d)).

Fig. 1. Captured video images of fiber fuse ignition and propagation along a single mode silica glass optical fiber pumped by a 9.0 W and 1.48 μm laser light. See the original video at http://www.youtube.com/watch?v=yjX5dU1EkTk (See also Table 3)

See http://pubman.mpdl.mpg.de/pubman/item/escidoc:1058545 and 1058546 for errata and a Japanese translation, respectively.

important milestones in optical communication, including the ultimate loss reduction of silica glass optical fibers by Kanamori et al. (1986) and the invention of Er-doped fiber amplifiers by Mears et al. (1987). Although this sensational phenomenon must have attracted some interest since its discovery, the number of related papers increased only after the turn of the century as a result of the rapid development of high power light sources (see Fig. 2).

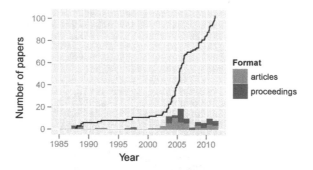

Fig. 2. Presentations and publications on the fiber fuse phenomenon. Total number (line) and yearly increments (bars). Source: http://www.geocities.jp/tokyo_1406/node3.html (See also Table 4)

Most of these researches were conducted from practical standpoints and have included techniques for the prevention and termination of a fiber fuse. In addition, studies from theoretical and/or microscopic viewpoints have been undertaken since 2003 (see Fig. 3). This paper briefly summarizes recent studies of macroscopic fiber fuse propagation where the fuse is regarded as a point without internal structures (Section 2) and the microscopic behavior of traveling plasma in relation to periodic void formation (Section 3). All the descriptions relate to silica-based fibers unless otherwise specified.

Fig. 3. Classification of fiber fuse studies with some key words and researchers.

2. Macroscopic behavior

2.1 Fiber fuse initiation, propagation and damage

Optical fibers were primarily developed to be transparent so that they would transmit light as far as possible. Therefore, at first it seems strange that transmitting light can actually destroy an optical fiber. However, it becomes reasonable when we consider that the absorption of the waveguide materials increases at elevated temperatures. Heat is the key to fiber fuse initiation.

Kashyap (1988) reported a steep increase in absorption over 1050 °C through a one-meter-long Ge-doped single-mode silica glass fiber (see Fig. 4). Shuto et al. (2004a) suggested three factors as possible origins of this absorption: (i) point-defect (Ge E' center) formation (see Eqs. (1) and (2)), (ii) electronic conductivity due to the thermal ionization of a Ge-doped silica core (Eq. (2)), and (iii) the thermochemical production of SiO in silica glass (Eq. (3)). Davis et al. (1996) reported that the threshold power density required for fiber fuse initiation was of the order of 3 MW/cm^{-1} among typical single mode fibers operated at 1.06 μm regardless of fiber type or core composition.

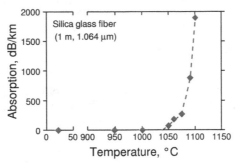

Fig. 4. Temperature dependence of the absorption in a single-mode Ge-doped silica glass fiber reported by Kashyap (1988).

$$-\overset{|}{\underset{/}{\text{Ge}}} -\text{O} - \overset{/}{\underset{\backslash}{\text{Si}}} - \longrightarrow -\overset{|}{\underset{/}{\text{Ge}}} - \overset{/}{\underset{\backslash}{\text{Si}}} - + \frac{1}{2}\text{O}_2 \uparrow \tag{1}$$

$$-\overset{|}{\underset{/}{\text{Ge}}} - \overset{/}{\underset{\backslash}{\text{Si}}} - \longrightarrow -\overset{|}{\underset{/}{\text{Ge}}} \,{}^\bullet\; \overset{/}{\underset{\backslash}{\text{Si}}}{}^{+} - + e^- \tag{2}$$

$$\text{SiO}_2 \longrightarrow \text{SiO} + \frac{1}{2}\text{O}_2 \uparrow \tag{3}$$

$$\text{SiO} \longrightarrow \text{Si} + \text{O} \longrightarrow \text{Si}^+ + \text{O}^+ + 2e^- \tag{4}$$

Another seemingly strange characteristic of a fiber fuse is the propagation of a bright spot toward the light source. However, this is intuitively understandable if we compare it with a grassfire (see Fig. 5). A solitary wave of fire or plasma persists as an irreversible reaction region being fueled from the front, emitting light and heat all around, and leaving cinders or damage behind it. Such a solitary wave can be considered a "dissipative soliton" as discussed later.

In the fiber fuse reaction region, ionized gas plasma is enclosed by a molten glass layer, and this situation is maintained by the pump laser energy. Shuto (2010) mentioned that most of the plasma is generated through the reaction described by Eq. (4). The temperature is estimated to be more than a few thousand degrees Kelvin by Hand & St. J. Russell (1988) and Dianov, Fortov, Bufetov, Efremov, Rakitin, Melkumov, Kulish & Frolov (2006).

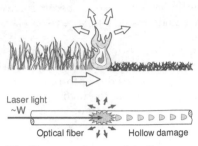

(a) Grassfire at Mt. Wakakusayama, a UNESCO World Heritage site, before the onset of spring in Nara, Japan. ©Yoshitaka Inoue

(b) Energy flows into/from a propagating reaction zone indicated with arrows.

Fig. 5. Comparison between grassfire and fiber fuse.

The propagation speed and pump power dependence of a fiber fuse have been investigated by many researchers. The speed is generally of the order of meters per second and increases almost linearly with pump laser power. However, Davis et al. (1997) pointed out that the slope vs. power density varies for different types of fibers.

As for the threshold power needed for fuse propagation, P_{th}, Dianov, Bufetov, Frolov, Plotnichenko, Mashinskii, Churbanov & Snopatin (2002) pointed out that mode field diameter (MFD), $2r_s$, is the dominating factor, i. e., the threshold power density ($I_{th} = P_{th}/\pi r_s^2$) varies approximately inversely with the MFD for several types of fibers pumped at 1.06–1.48 μm (see also Bufetov & Dianov (2005)). On the other hand, Seo et al. (2003) reported a linear relation between P_{th} and MFD, which is parallel to the Dianov's relation. Table 1 lists the P_{th} values for some standard silica based fibers.

Fiber	P_{th}/W	λ/μm	P_{th}/W	λ/μm	
SMF	1.0	1.064	1.4	1.467	Seo et al. (2003)
			~1.2	1.48	Todoroki (2005c)
			1.39	1.55	Abedin & Morioka (2009)
DSF	1.2	1.064	0.65	1.467	Seo et al. (2003)
			~1.1	1.55	Abedin (2009)
DCF			~0.7	1.55	Abedin (2009)

DSF: Dispersion Shifted Fiber, DCF: Dispersion Compensating Fiber

Table 1. Threshold power of fiber fuse propagation, P_{th}, for various fibers. See also Fig. 6 in Takenaga, Omori, Goto, Tanigawa, Matsuo & Himeno (2008).

After the fiber fuse has passed, a hollow damage train is left behind, and the fiber is no longer able to guide light. Figure 6 shows an example of fuse damage; a fiber fuse propagated from right to left and terminated at the position shown by the arrow due to a gradual reduction in the pump power. The pair of horizontal lines surrounding these voids is the border of the region modified by the passage of the hot plasma. Its diameter is larger than the original core size seen on the left of the arrow and increases with the pump power.

Kashyap (1988) detected O_2 gas inside the voids using Raman microscopy. The adjacent glass layer is expected to be densified. Dianov et al. (1992) provided supporting evidence, namely that refractive index around the void increased after the passage of the fiber fuse and subsequently decreased after fiber annealing for several seconds at about 1000 °C.

Fig. 6. Fiber fuse damage train in a standard single mode fiber (Corning SMF-28). After the ignition of the fiber fuse, the pump power of the laser (1.48 μm) decreased from 7 W (c), 3.5 W (b) to ~1.2 W (a) until its self-extinction at the position shown by the arrow.

The top of the void train provides a strong scattering point if the light is launched again. Yamada et al. (2011) warned that this scattered light heats and burns the surrounding coating and nylon jacket.

2.2 Dissipative soliton and termination technology

If we wish to avoid a fiber fuse, we should consider the energy flow into and from the traveling plasma. A "dissipative soliton" is a useful concept for this purpose and is defined by Akhmediev & Ankiewicz (2005) as follows:

A dissipative soliton is a localized structure which exists for an extended period of time, even though parts of the structure experience gain and loss of energy and/or mass. $\cdots$ These solitons exist in "open" systems which are far from equilibrium.

Figure 7 compares a dissipative soliton and a conventional soliton that transfers energy and/or mass.

Fig. 7. Comparison of conventional and dissipative solitons.

On the basis of this energy flow, several researchers performed computer simulations of fiber fuse propagation and succeeded in reproducing the experimental propagation speed values. Their equations describing the energy flow are listed in Table 2. For the absorption coefficient, Rocha et al. (2009) used the Arrhenius form, $\alpha = \alpha_0 \exp(-T'/T)$, whereas other researchers used a custom function reproducing the steep temperature dependence shown in Fig. 4.

Shuto et al. (2004b) performed precise calculations using polar coordinates in which radiation heat loss appeared in the boundary condition. They also discussed the wavelength and power density dependence of the propagation speed. Akhmediev et al. (2008) described fiber fuse propagation as a "dissipative soliton" on the basis of a simple one dimensional calculation. Rocha et al. (2009) introduced a radiation loss term from the fiber surface and simulated three different types of fibers using MFD and α_0 as parameters (see also Rocha et al. (2010)).

The dissipative soliton disappears when the supply of energy or matter is reduced, or the radiant heat and/or light is increased so that the system parameters move outside the range in which the soliton can exist. With a fiber fuse, it stops when the pumping power falls below the threshold power, P_{th}, or when the energy removal is increased. This is the design principal behind "fiber fuse terminators", which have been proposed by many researchers.

	Light-induced heat	Diffusion	Radiation
Shuto et al. (2004b), Golyatina et al. (2004) : $\rho C_p \dfrac{\partial T}{\partial t} =$	$\alpha(T)I$	$+\kappa \left(\dfrac{\partial^2 T}{\partial z^2} + \dfrac{\partial^2 T}{\partial r^2} + \dfrac{1}{r}\dfrac{\partial T}{\partial r} \right)$	
Akhmediev et al. (2008) : $\dfrac{\partial T}{\partial t} =$	$\chi\alpha(T)I$	$+D\dfrac{\partial^2 T}{\partial z^2}$	$-k(T - T_0)$
Rocha et al. (2009) : $\rho C_p \dfrac{\partial T}{\partial t} =$	$\alpha_{\mathrm{Arr}}I$	$+\kappa\dfrac{\partial^2 T}{\partial z^2}$	$-\sigma_s\epsilon_e(T^4 - T_0^4)$

Table 2. Equations used for simulating fiber fuse propagation. ρ: density, C_p: specific heat, T: plasma temperature, T_0: environment temperature, $\alpha(T)$ and α_{Arr}: absorption coefficient (see text), κ: thermal conductivity, σ_s: Stefan-Boltzmann constant, ϵ_e: surface emissivity.

Hand & Birks (1989) demonstrated a fiber fuse termination at a special segment where the waveguide structure was modified to expand the pump laser beam (see Fig. 8 (a)). This is because a reduced power density in the core region exhausts the feed for the plasma. Later, Yanagi et al. (2003) developed a detachable device for practical use. However, we should note that this device cannot capture the fuse if the pump power exceeds the assumed limit.

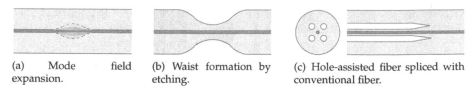

(a) Mode field expansion.

(b) Waist formation by etching.

(c) Hole-assisted fiber spliced with conventional fiber.

Fig. 8. Waveguide structures for in-line fiber fuse terminators.

Dianov, Bufetov & Frolov (2004) proposed a different structure in which the thickness of the cladding layer is reduced by chemical etching as illustrated in Fig. 8 (b). The fiber fuse is arrested here because its internal pressure cannot be maintained at this weakened segment and this causes deformation. Takenaga, Tanigawa, Matsuo, Fujimaki & Tsuchiya (2008) discovered that fiber fuse propagation is not as easy in hole-assisted fibers (HAFs) as in conventional fibers. The termination occurs near the splice point of these fibers (see Fig. 8 (c)) and the penetration length into the HAF side increases as the pump power decreases. Thus, it is reasonable to consider that the holes release the internal pressure of the plasma to destabilize it. This behavior was directly observed by Hanzawa et al. (2010) with an ultra-high speed video camera. Recent proposals for devices using holey fibers are discussed in section 2.3.

Another approach to fiber fuse termination is to interrupt the light source after detecting a propagating fiber fuse. Abedin et al. (2009) proposed a remote detection method that employs a backreflected light from the void train (see Fig. 9 (a)). They found that characteristic signals appeared in optical coherence-domain reflectometry after the ignition of the fiber fuse.

Rocha et al. (2011) demonstrated a local detection technique that uses a fiber Bragg grating as a temperature sensor placed in thermal contact with the propagation line (see Fig. 9 (b)). A heat pulse of a few degrees centigrade was detected as an increase in the Bragg wavelength monitored with an optical interrogator.

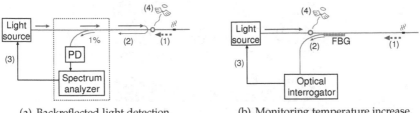

(a) Backreflected light detection.   (b) Monitoring temperature increase.

Fig. 9. Schematic diagram of fiber fuse terminator. (1) propagation after ignition, (2) signal detection, (3) interruption of the light source and (4) termination.

2.3 Fiber fuse under special conditions

A fiber fuse can be initiated not only with a CW laser but also with a pulsed laser. Kashyap (1988) reported that mode locking has little effect on the velocity of damage propagation, but does alter the shape of the periodic voids. Dianov, Fortov, Bufetov, Efremov, Frolov, Schelev & Lozovoi (2006) demonstrated a detonation-like propagation mode with velocities up to 3 km/s that appeared when a Q-switched laser was used for pumping (wavelength: 1.064 μm, pulse repetition frequency: 5 kHz, pulse duration: 250 ns, pulse energy: up to 0.6 mJ, and maximum pulse power: 3.0 kW (3.8 W average)). Traveling plasma appeared only for a pulse duration and the damage included large cracks with a diameter up to 120 μm.

There have been several reports on non-conventional fibers. Lee et al. (2006) investigated fuse propagation over polarization-maintaining fibers and found that the power threshold for fast axis alignment is larger than that for slow axis alignment. Wang et al. (2008) showed some damage photographs of crack propagation in double-clad fiber for high power use.

Photonic crystal fibers (PCFs) also allow fiber fuse propagation. However, Dianov, Bufetov, Frolov, Chamorovsky, Ivanov & Vorobjev (2004) demonstrated that their threshold power is approximately ten times higher than that for conventional fibers. The reason is apparently the same as that discussed for HAFs above; the air holes reduce the plasma density. Thus, it is possible to use this fiber to make a fiber fuse terminator. Kurokawa & Hanzawa (2011) observed a fuse termination in situ at a splice point between PCF and conventional fiber. Ha et al. (2011) proposed another terminator using a hollow optical fiber.

Dianov, Bufetov, Frolov, Mashinskii, Plotnichenko, Churbanov & Snopatin (2002) reported the destruction of chalcogenide and fluoride glass fibers pumped at less than 1 W. They observed a distraction wave without plasma that thermally decomposed the entire cross section of the fiber. This is because these materials decompose at a much lower temperature than silica glass.

3. Microscopic behavior

Another characteristic behavior of a fiber fuse is the formation of periodic bullet-shaped voids. These voids are formed just after the passage of running plasma, and this has been experimentally confirmed by ultra-high speed photography (Bufetov & Dianov (2005); Bufetov et al. (2005), or see Fig. 10).

Two ideas have been proposed for the driving force behind this void formation. Atkins et al. (2003) suggested that Rayleigh instability, or positive surface tension, minimizes the interface area between plasma and molten glass following the analogy of falling water droplets and an air jet in a fluid (see Fig. 11). On the other hand, Yakovlenko (2004) pointed out that this idea has certain inadequacies, that is, it overlooks the high viscosity of molten silica and the

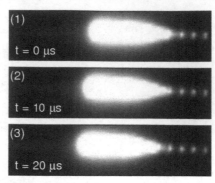

Fig. 10. Ultra-high speed photographs of fiber fuse propagation pumped with 9 W 1.48 μm light (Todoroki (2005b)). Discrete scattering points from the periodic voids (22 μm-interval) are clearly seen after the running plasma.

disappearance of surface tension at the elevated temperatures. Instead, he mentioned the electrostatic repulsion between negative charge layers induced at the plasma-liquid interface (see also Yakovlenko (2006a)). However, neither idea can explain why the voids look like bullets. Although Yakovlenko (2006b) pointed out that a bullet shape appeared in his simulated temperature profile of a running plasma, there is no description of its periodic appearance in the time domain.

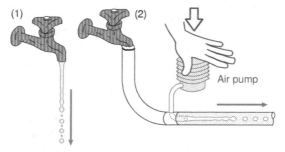

Fig. 11. Examples of jet breakdown due to Rayleigh instability, (1) A jet of water broken up into droplets (see Isenberg (1992) p. 131) and (2) a bubble train in a water flow (see Chandrasekhar (1981) p. 540).

Meanwhile, I have been investigating the mechanism of periodic void formation through an experiment-based approach, namely, a morphological analysis of void shapes and ultra-high speed photography. Section 3.1 discusses the requirements for periodic void formation. Then, the quench-induced deformation of the hollow melt is clarified by a statistical analysis of damage photographs (Section 3.2). Finally, the mechanism of bullet-like void formation is proposed (Section 3.3). All the discussion relates to silica-based step-indexed single-mode optical fibers unless otherwise specified.

3.1 Pump power dependence of hollow damage morphology

We should note that not all the fused damages has the appearance of periodic bullets. Figure 12 shows the front part of a fused damage train that remained in Corning SMF-28 fibers after the light sources had been turned off (1.2–9 W, 1.48 μm, CW; Todoroki (2005c)). Bullet-shaped periodic voids appear when the pump power exceeds 2 W (see (a)–(d)) or ~1.3 W (g) near the propagation threshold. The latter mode is very unsteady with respect to fluctuations in laser power and/or waveguide structure, i. e. the periodicity disappears easily and sometimes the fuse vanishes[1]. For the former stable mode, the interval of the periodic voids increases with the pump power and a long and narrow void is left at the top of the damage train.

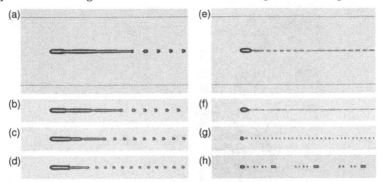

Fig. 12. Optical micrographs showing the front part of the fiber fuse damage generated in single-mode silica glass fibers (Corning SMF-28). The powers of the pump laser (1.48 μm) are (a) 9.0 W, (b) 7.0 W, (c) 5.0 W, (d) 3.5 W, (e) 2.0 W, (f) 1.5 W, (g) ~1.3 W, and (h) ~1.2 W. The thin two lines at the top and bottom of (a) and (e) are the edges of the 125 μm diameter fiber. (Todoroki (2005c))

These top voids are valuable evidence for exploring the state of traveling plasma. In fact, these photographs show good agreement with the in-situ image of fiber fuse propagation shown in Fig. 13. In both cases, the shape along the axial direction is asymmetric when the pump power exceeds 2 W. Thus, this asymmetric shape is expected to be the origin of the stable periodic void formation. However, these in-situ images provide no further information about the void formation process owing to their poor resolution. Thus, the shape of the damage sites are analyzed instead in the next subsection.

3.2 Deformation of hollow melt during quenching

The period of one void formation is estimated to be a few tens of microseconds based on the propagation speed and void interval. For example, the periodic voids shown in Fig. 12 (a)–(d) were generated every 18.7 μs (9 W) – 25.4 μs (3.5 W). During this period, the plasma was found to propagate at a constant speed (Todoroki (2005a)). Since the damage structure varies according to the moment at which the light source is turned off during this cycle, at least 40 samples were prepared in order to collect a variety of damage patterns.

In addition, to maximize the quenching rate of the melt, a fiber fuse was terminated in a segment where the colored nylon jacket (0.9 mmϕ) over the cladding had been removed beforehand (about 20 cm at the maximum). This is because color pigments in the nylon jacket scatters the visible radiation from the inside and the backscattered light is re-absorbed by the melt with absorptive species shown in Eqs. (1)–(3) to generate heat. This light-heat conversion

---

[1] For example, see the video in Fig. 2 of Todoroki (2005d) in which the fuse stopped after it passed through a splicing point in a jacket-free segment.

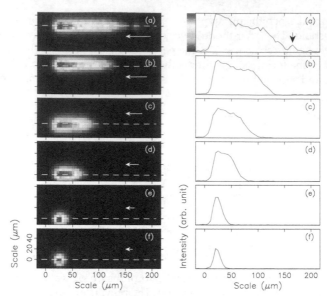

Fig. 13. Ultrahigh-speed photograph of visible light emission from a fiber fuse (left, gray-scale image is converted to colored-scale image, exposure time: 1 μs) and intensity profiles along the dashed line in each image (right). The powers of the pump laser (1.48 μm) are (a) 9.0 W, (b) 7.0 W, (c) 5.0 W, (d) 3.5 W, (e) 2.0 W, and (f) 1.5 W. Each horizontal arrow indicates the distance that the plasma moves in 40 μs. (Todoroki (2005c))

occurs throughout the fiber fuse propagation but is absent in the bare fiber segment just before the fiber fuse termination[2].

To compare the damage sites precisely, the void size should be properly normalized because it varies sensitively with the pump power. For example, the interval of the periodic voids, Λ, among the 81 samples pumped with 9 W 1.48 μm light ranged from 21.7 μm to 22.7 μm. This must be due to the fluctuation of pump laser power and the loss of the fiber between the fuse and the light source. Thus, the following two parameters are defined on the basis of Λ,

$$x_1' = x_1/\Lambda, \qquad l_2' = l_2/\Lambda \qquad (5)$$

where x_1 is a parameter describing the top position of the first void in a virtual scale graduated on Λ (see the vertical lines in Fig. 14) and l_2 is the length of the second void.

The left column in Fig. 15 ((1)–(6)) shows selected photographs of the samples pumped with 9 W 1.48 μm light. They are sorted in order of increasing x_1', with the intention of rearranging them in chronological order within the void formation cycle (Todoroki (2005a;c)). The sorted photographs suggest that l_2' has a tendency to decrease with increasing x_1' (see the inset table). This correlation among all the 40 samples is shown on the left in Fig. 16.

The sequence of the photographs seems to capture the moment at which the long top void is divided in two by a melt bridge (Todoroki (2005a;c)). However, we should not overlook a possibility that these structures are the result of the modification that occurred during the fiber

---

[2] The plasma decay time in a bare fiber is reported to be less than 7 μs (Todoroki (2005a)). Detailed description of this light-heat conversion will be published soon (Todoroki (2011)).

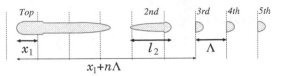

Fig. 14. Definition of the parameters used in Eq. (5). The vertical lines are placed at the bottom of bullet-like voids at intervals of Λ. Then, the distance between the top of the first void and periodic voids is written as $x_1 + n\Lambda$ (n: natural number).

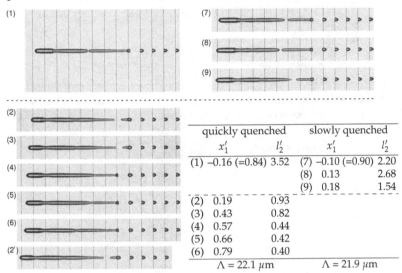

quickly quenched		slowly quenched	
x_1'	l_2'	x_1'	l_2'
(1) –0.16 (=0.84) 3.52		(7) –0.10 (=0.90) 2.20	
		(8) 0.13	2.68
		(9) 0.18	1.54
(2) 0.19	0.93		
(3) 0.43	0.82		
(4) 0.57	0.44		
(5) 0.66	0.42		
(6) 0.79	0.40		
$\Lambda = 22.1\ \mu m$		$\Lambda = 21.9\ \mu m$	

Fig. 15. Optical micrographs of the damage sites pumped with 9 W 1.48 μm light. (1) – (6): samples whose quenching point has no plastic coating, and (7) – (9): with coating (see text). The photograph at the bottom (2') is the same as that at the top (2), shifted 22.1 μm to the left. The inset table lists the size parameters defined in Eq. (5), that are plotted as red points in Figs. 16 and Fig. 17.

fuse quenching period. Fortunately, the histogram of l_2' shown in the right of Fig. 16 provides a clue to this problem. That is, the l_2' distribution is strongly biased below 1.0. In other words, void structures like (1) shown in Fig. 15 ($l_2' > 1.0$) appeared less frequently than (2) – (6). If these void structures are stable during quenching, the l_2' distribution is expected to be independent on the quenching rate. Thus, another set of the samples was prepared in which a fiber fuse was quenched in a colored nylon jacketed segment with an outer diameter of 0.9 mm. Hereafter, they are referred to as 'slowly quenched' whereas the previous samples were referred to as 'quickly quenched'. The result shown in Fig. 17 is clearly different from that shown in Fig. 16; the correlation between x_1' and l_2' became considerably weaker and the l_2' histogram became relatively flat.

This behavior is well explained by an assumption that the melt surrounding the plasma tends to form a bridge inside the cavity after the plasma has been extinguished and before the melt is frozen (see the lower white arrow in Fig. 18 between (3-a) and (c)). In other words, after the quench begins, a melt with a long hollow space ($l_2' < 1$) becomes unstable and forms a bridge inside the cavity ($l_2' > 1$). In this case, the bridge position is influenced very little by

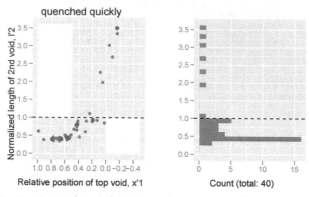

Fig. 16. Correlation between x_1' and l_2' (left) and histogram of l_2' (right) for the samples pumped with 9W 1.48 μm light in which the fiber fuse was extinguished in a coating-free segment (see text). The red points are the results for samples (1) – (6) shown in Fig. 15.

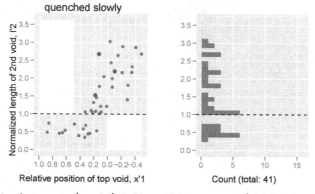

Fig. 17. Correlation between x_1' and l_2' (left) and histogram of l_2' (right) for the samples pumped with 9W 1.48 μm light in which the fiber fuse was extinguished in a coated segment (see text). The red points are the results for samples for (7) – (9) shown in Fig. 15.

the moment at which the power is turned off, and there is little correlation between x_1' and l_2' in the resulting structures. In addition, the bridge width in the slowly quenched samples is larger than that of the quickly quenched samples (compare Fig. 15 (7) – (9) and (1)). This suggests that bridge growth was promoted by slow quenching.

Possible driving forces for this bridge formation are, firstly, the sudden pressure/temperature decrease in the hollow cavity that occurs as a result of the laser power being switched off and the subsequent deposition of quenched gas, and secondly, the negative surface tension of glass melt (Yakovlenko (2006b)). The temperature dependence of the surface tension is given by Eötvös formula, $\gamma(T) = k(\rho/M)^{2/3}(T_{cr} - T)$, where k is a constant, ρ density, M molar weight, and T_{cr} critical temperature. Although the experimental data for the silica melt are not available, Yakovlenko (2006b) mentioned that the tension is expected to be negative at elevated temperatures (more than $\sim$3000 K) and surface creation promoted.

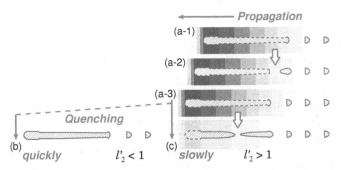

Fig. 18. Model of void formation during propagation and quenching of fiber fuse. Bridge formation is shown by white arrows.

The bridge that remained in the quickly quenched samples with $l'_2 > 1$ as in Fig. 15 (1) were probably formed after the power had been turned off. The low probability of the samples with $l'_2 > 1$ shown on the right in Fig. 16 supports this hypothesis. Moreover, this is confirmed by the recent result that I obtained by the in-situ videography of fiber fuse propagation shown in the left of Fig. 19. If there is a bridge before quenching, i. e. during propagation, no light emission is expected from the bridge. However, the centers of the intensity profiles have no dark regions.

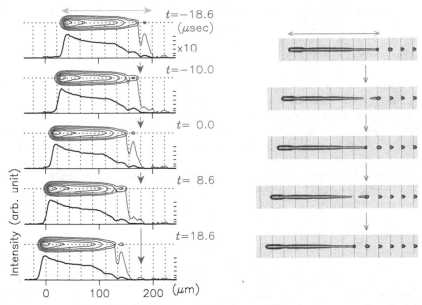

Fig. 19. Intensity profiles of visible light emission from a fiber fuse propagating through a single-mode optical fiber pumped by 9 W 1.48 μm light (left, exposure time: 0.37 μs) and photographs of hollow damage for comparison, generated under the same condition (right). The vertical line interval is 22 μm. The vertical dashed lines are placed on the scattering points from the periodic voids. The photographs are the same as those shown in Fig. 15 (2) and (4).

This is more clearly recognized when we compare these profiles with the void photographs as shown on the right. These photographs are arranged to coincide with the in situ images with respect to the following two points; (1) the full width of the profile matches the length of the top void (see the green arrows on the top) and (2) the weak scattering points on the vertical dashed line match the top of the periodic voids on the vertical solid line. The blue arrows are reference points for this comparison.

Consequently, the melt surrounding the traveling plasma tends to form a bridge inside the cavity after the light source has been turned off. However, this action is suppressed by fast quenching (see Fig. 18 (b)). Therefore, the damage photographs of quickly quenched samples without such a bridge (see Fig. 15 (2) – (6)) constitute useful data for discussing the periodic process of fiber fuse propagation in the next subsection.

3.3 Bridge formation during fiber fuse propagation

According to the sequence in Fig. 15 (2) – (6), the bridge formation process occurs between (6) and (2′). The hottest region of the plasma passed this position $\sim$120 μs earlier and the temperature is decreasing. Its quenching rate is much slower than that induced by suddenly turning off the light source because the hot plasma is still alive nearby and is moving away at about 1 m/s (= 1 μm/μs). Thus, this situation, illustrated in Fig. 18 (a-1) and (a-2), is almost the same as the slow quenching case (see the lower white arrow) except that the bridge is compressed by the hot plasma.

This compression explains why the periodic voids look like bullets. This process is frozen in the photographs shown in Fig. 15 (2) – (6). After a bridge appears, the detached void begins to shrink due to the pressure from the hot plasma until the surrounding melt solidifies. Since the rear side of the new void solidifies earlier than the front, the rear shape remains round whereas the front becomes flatter.

It is interesting to find that an early sign of bridge formation is recorded in the in-situ observation result shown on the left in Fig. 19. A weak modulation appeared on the tail of the light emission profile and its interval is the same as that of the periodic voids. However, it appeared only when the pump power was 9 W (see the black arrow in Fig. 13). This modulation may suggest the instability of the plasma or the surrounding melt as proposed by Atkins et al. (2003) (Rayleigh instability) and Yakovlenko (2004) (induced electrostatic repulsion) but further study is needed.

The void formation sequence can be modified by controlling certain external conditions. Bufetov et al. (2008) observed a large-scale periodic void train in an optical fiber that allows the interference of the LP_{01} and LP_{02} modes. Figure 20 shows the void train surrounded by a region of modified refractive index. Its interval was found to coincide with that of the interference pattern.

Todoroki (2008) reported the breakage of the periodic void pattern over hetero-core splice points as shown in Fig. 21. Traveling plasma temporarily stopped forming voids when the core was expanded from HI 1600 to SMF-28e, whereas it left some long voids when the core size was reduced. It is interesting to find a similar tendency in Bufetov's case at the inflection points of the interference. That is, the void train disappears at the segment where the mode field is increasing (see dashed lines b in Fig. 20), and a long void appears where the mode field is reduced (see dashed lines a). Moreover, some of the voids have a flat area whose direction is opposite to that of the regular bullet-like voids (see arrows in Figs. 20 and 21). This behavior must be due to the modulated internal pressure of the plasma. To clarify the mechanism,

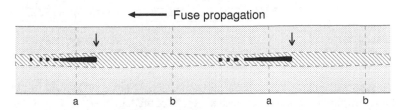

Fig. 20. Illustration of large-scale periodic void train left in an optical fiber allowing the interference of two propagation modes of 1.07 μm light. This is reproduced from a photograph in Bufetov et al. (2007). The diameter of the fiber cladding is 125 μm. The vertical dashed lines are on the inflection points of the modified refractive-index area shown as a red hatched region.

further investigation is needed including the in-situ observation of traveling plasma and a statistical analysis of quickly quenched void samples.

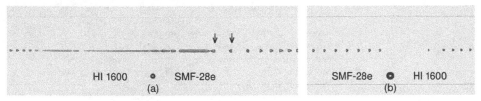

Fig. 21. Photographs of fiber fuse damage over the splicing point between HI 1060 and SMF-28e (Corning). The hollow spheres in the cladding are located at the border between two fibers and were accidentally captured during splicing. The fuse was initiated with 9W 1.07 μm light that propagates in a multi-mode through SMF-28e.

4. Summary

Knowledge accumulated about fiber fuse propagation since 1988 is briefly summarized. From a macroscopic viewpoint, the dissipative soliton concept and an analogy with grassfire help us understand this strange phenomenon. The strong heat-induced absorption of silica glass and the highly confined supply of laser energy cause captured plasma to shift to the light source along the fiber leaving catastrophic damage behind it. From a microscopic viewpoint, the periodic void formation process was unveiled by the statistical analysis of void shapes and ultra-fast videography. The bullet-like shape of the damage train results from the formation of an intrinsic bridge inside the hollow silica melt behind the traveling plasma and the successive compression of detached voids under a steep temperature gradient along the fiber.

5. Acknowledgements

I am grateful to Mr. Keisuke Aizawa, Mr. Arata Mihara and Mr. Yousuke Suzuki (Photron Ltd.) for helping with the ultrahigh-speed videography experiment.

	Todoroki (2005c)	Todoroki (2005d)	Todoroki (2008)
Ignition:	—	Fig. 3*	—
Propagation:	Fig. 1, Fig. 3*	—	Movie S1* & S2*
Self-termination:	—	Fig. 2, Fig. 7*	—
Void formation model:†	Fig. 7	—	—

Table 3. Fiber fuse video clips available on the net. *: Ultra-high speed video, †: outdated because all the samples were slowly quenched (see Section 3.2).

	Russia	Japan
Kashyap (1988)		
Kashyap & Blow (1988)		
Hand & St. J. Russell (1988)		
Hand & Birks (1989)		
Davis et al. (1996)	Dianov et al. (1992)	
Davis et al. (1997)		
	Dianov, Bufetov, Frolov, Plotnichenko, Mashinskii, Churbanov & Snopatin (2002)	*Japan*
Atkins et al. (2003)	Dianov, Bufetov, Frolov, Mashinskii, Plotnichenko, Churbanov & Snopatin (2002)	Yanagi et al. (2003)
		Seo et al. (2003)
	Dianov, Bufetov, Frolov, Chamorovsky, Ivanov & Vorobjev (2004)	
	Dianov, Bufetov & Frolov (2004)	*Shuto et al. (2004a)*
	Yakovlenko (2004)	*Shuto et al. (2004b)*
	Golyatina et al. (2004)	
	Bufetov & Dianov (2005)	Todoroki (2005b)
	Bufetov et al. (2005)	Todoroki (2005c)
		Todoroki (2005a)
		Todoroki (2005d)
Lee et al. (2006)	Dianov, Fortov, Bufetov, Efremov, Frolov, Schelev & Lozovoi (2006)	
	Dianov, Fortov, Bufetov, Efremov, Rakitin, Melkumov, Kulish & Frolov (2006)	
	Yakovlenko (2006a)	
	Yakovlenko (2006b)	
	Bufetov et al. (2007)	
	Bufetov et al. (2008)	Takenaga, Omori, Goto, Tanigawa, Matsuo & Himeno (2008)
Akhmediev et al. (2008)		Takenaga, Tanigawa, Matsuo, Fujimaki & Tsuchiya (2008)
Wang et al. (2008)		Todoroki (2008)
		Abedin & Morioka (2009)
		Abedin et al. (2009)
Rocha et al. (2009)		Abedin (2009)
		Shuto (2010)
Rocha et al. (2010)		Hanzawa et al. (2010)
Ha et al. (2011)		Kurokawa & Hanzawa (2011)
Rocha et al. (2011)		Yamada et al. (2011)
		Todoroki (2011)

Table 4. Chronological table of fiber fuse research cited in this paper. Theoretical works are given in italics.

6. References

Abedin, K. S. (2009). Remote sensing of fiber fuse propagation using RF detection, *The Technical Report of The Proceedings of The Institute of Electronics, Information and Communication Engineers, OPE, Optoelectronics* 109(159): 43–46.

Abedin, K. S. & Morioka, T. (2009). Remote detection of fiber fuse propagating in optical fibers, *Proceedings of Optical Fiber Communication/National Fiber Optic Engineers Conference.* (OThD5).

Abedin, K. S., Nakazawa, M. & Miyazaki, T. (2009). Backreflected radiation due to a propagating fiber fuse, *Optics Express* 17(8): 6525–6531.

Akhmediev, N. & Ankiewicz, A. (2005). Dissipative solitons in the complex Ginzburg-Landau and Swift-Hohenberg equations, *Dissipative Solitons*, Vol. 661 of *Lecture Notes in Physics*, Springer-Verlag, Berlin, pp. 1–17.

Akhmediev, N., St. J. Russell, P., Taki, M. & Soto-Crespo, J. M. (2008). Heat dissipative solitons in optical fibers, *Physics Letters A* 372(9): 1531–1534.

Atkins, R. M., Simpkins, P. G. & Yablon, A. D. (2003). Track of a fiber fuse: a Rayleigh instability in optical waveguides, *Opt. Lett.* 28(12): 974–976.

Bufetov, I. A. & Dianov, E. M. (2005). Optical discharge in optical fibers, *Physics-Uspekhi* 48(1): 91–94.

Bufetov, I. A., Frolov, A. A., Dianov, E. M., Fortov, V. E. & Efremov, V. P. (2005). Dynamics of fiber fuse propagation, *Optical Fiber Communication Conference, 2005. Technical Digest. OFC/NFOEC*, Vol. 4, Anaheim, CA. (OThQ7).

Bufetov, I. A., Frolov, A. A., Shubin, A. V., Likhachev, M. E., Lavrishchev, C. V. & Dianov, E. M. (2007). Fiber fuse effect: New results on the fiber damage structure, *Proceedings of the 33rd European Conference on Optical Communication*, Vol. 1, IEE's Photonics Professional Network, Berlin, Germany, pp. 79–80. (Mon 1.5.2).

Bufetov, I. A., Frolov, A. A., Shubin, A. V., Likhachev, M. E., Lavrishchev, S. V. & Dianov, E. M. (2008). Propagation of an optical discharge through optical fibres upon interference of modes, *Quantum Electronics* 38(5): 441–444.

Chandrasekhar, S. (1981). *Hydrodynamic and Hydromagnetic Stability*, International Series of Monographs on Physics (Oxford, England), Dover Publications.

Davis, D. D., Mettler, S. C. & DiGiovani, D. J. (1996). Experimental data on the fiber fuse, *in* H. E. Bennett, A. H. Guenther, M. R. Kozlowski, B. E. Newnam & M. J. Soileau (eds), *27th Annual Boulder Damage Symposium: Laser-Induced Damage in Optical Materials: 1995*, Vol. 2714 of *SPIE Proceedings*, SPIE, pp. 202–210. (Boulder, CO, USA, 30 Oct. 1995).

Davis, D. D., Mettler, S. C. & DiGiovani, D. J. (1997). A comparative evaluation of fiber fuse models, *in* H. E. Bennett, A. H. Guenther, M. R. Kozlowski, B. E. Newnam & M. J. Soileau (eds), *Laser-Induced Damage in Optical Materials: 1996*, Vol. 2966 of *SPIE Proceedings*, SPIE, pp. 592–606. (Boulder, CO, USA, 7 Oct 1996).

Dianov, E. M., Bufetov, I. A. & Frolov, A. A. (2004). Destruction of silica fiber cladding by the fuse effect, *Opt. Lett.* 29(16): 1852–1854.

Dianov, E. M., Bufetov, I. A., Frolov, A. A., Chamorovsky, Y. K., Ivanov, G. A. & Vorobjev, I. L. (2004). Fiber fuse effect in microstructured fibers, *IEEE Photon. Technol. Lett.* 16(1): 180–181.

Dianov, E. M., Bufetov, I. A., Frolov, A. A., Mashinskii, V. M., Plotnichenko, V. G., Churbanov, M. F. & Snopatin, G. E. (2002). Catastrophic destruction of fluoride and chalcogenide optical fibers, *Electron. Letters* 38(15): 783–784.

Dianov, E. M., Bufetov, I. A., Frolov, A. A., Plotnichenko, V. G., Mashinskii, V. M., Churbanov, M. F. & Snopatin, G. E. (2002). Catastrophic destruction of optical fibres of various composition caused by laser radiation, *Quantum Electron.* 32(6): 476–478.

Dianov, E. M., Fortov, V. E., Bufetov, I. A., Efremov, V. P., Frolov, A. A., Schelev, M. Y. & Lozovoi, V. I. (2006). Detonation-like mode of the destruction of optical fibers under intense laser radiation, *J. Exp. Theo. Phys. Lett.* 83(2): 75–78.

Dianov, E. M., Fortov, V. E., Bufetov, I. A., Efremov, V. P., Rakitin, A. E., Melkumov, M. A., Kulish, M. I. & Frolov, A. A. (2006). High-speed photography, spectra, and temperature of optical discharge in silica-based fibers, *IEEE Photon. Technol. Lett.* 18(6): 752–754.

Dianov, E. M., Mashinskii, V. M., Myzina, V. A., Sidorin, Y. S., Streltsov, A. M. & Chickolini, A. V. (1992). Change of refractive index profile in the process of laser-induced fiber damage, *Sov. Lightwave Commun.* 2: 293–299.

Golyatina, R. I., Tkachev, A. N. & Yakovlenko, S. I. (2004). Calculation of velocity and threshold for a thermal wave of laser radiation absorption in a fiber optic waveguide based on the two-dimensional nonstationary heat conduction equation, *Laser Physics* 14(11): 1429–1433.

Ha, W., Jeong, Y. & Oh, K. (2011). Fiber fuse effect in hollow optical fibers, *Opt. Lett.* 36(9): 1536–1538.

Hand, D. P. & Birks, T. A. (1989). Single-mode tapers as 'fibre fuse' damage circuit-breakers, *Electron. Lett.* 25(1): 33–34.

Hand, D. P. & St. J. Russell, P. (1988). Solitary thermal shock waves and optical damage in optical fibers: the fiber fuse, *Opt. Lett.* 13(9): 767–769.

Hanzawa, N., Kurokawa, K., Tsujikawa, K., Matsui, T., Nakajima, K., Tomita, S. & Tsubokawa, M. (2010). Suppression of fiber fuse propagation in hole assisted fiber and photonic crystal fiber, *J. Lightwave Technology* 28(15): 2115–2120.

Isenberg, C. (1992). *The Science of Soap Films and Soap Bubbles*, new edn, Dover Publications.

Kanamori, H., Yokota, H., Tanaka, G., Watanabe, M., Ishiguro, Y., Yoshida, I., Kakii, T., Itoh, S., Asano, Y. & Tanaka, S. (1986). Transmission characteristics and reliability of pure-silica-core single-mode fibers, *J. Lightwave Technology* 4(8): 1144–1150.

Kashyap, R. (1988). Self-propelled self-focusing damage in optical fibres, *Lasers '87; Proc. the Tenth Int. Conf. Lasers and Applications*, STS Press, McLean, VA, pp. 859–866. (Lake Tahoe, NV, Dec. 7-11, 1987).

Kashyap, R. & Blow, K. J. (1988). Observation of catastrophic self-propelled self-focusing in optical fibres, *Electron. Lett.* 24: 47–49.

Kurokawa, K. & Hanzawa, N. (2011). Fiber fuse propagation and its suppression in hole-assisted fibers, *IEICE Transactions on Communications* E94.B(2): 384–391.

Lee, M. M., Roth, J. M., Ulmer, T. G. & Cryan, C. V. (2006). The fiber fuse phenomenon in polarization-maintaining fibers at $1.55\mu m$, *Proc. of the Conference on Lasers and Electro-Optics (CLEO)*. (JWB66).

Mears, R. J., Reekie, L., Jauncey, I. M. & Payne, D. N. (1987). Low-noise erbium-doped fibre amplifier operating at 1.54 μm, *Electronics Letters* 23(19): 1026–1028.

Rocha, A. M., Antunes, P. F. C., Domingues, M. F. F., Facão, M. & André, P. S. (2011). Detection of fiber fuse effect using FBG sensors, *IEEE Sensors Journal* 11(6): 1390 –1394.

Rocha, A. M., Facão, M. & André, P. S. (2010). Study of fiber fuse effect on different types of single mode optical fibers, *in* D. Faulkner (ed.), *NOC/OC&I 2010 Proceedings : 15th European Conference on Networks and Optical Communications and 5th Conference on Optical Cabling and Infrastructure (ISBN: 9789729341939)*, Universidade do Algarve, Faro-Algarve, Portugal, pp. 71–75. (Presented on June 10).

Rocha, A. M., Facão, M., Martins, A. & André, P. S. (2009). Simulation of fiber fuse effect propagation, *International Conf. on Transparent Networks – Mediterranean Winter, 2009*, Angers, France. (FrP.12).

Seo, K., Nishimura, N., Shiino, M., Yuguchi, R. & Sasaki, H. (2003). Evaluation of high-power endurance in optical fiber links, *Furukawa Review* (24): 17–22.

Shuto, Y. (2010). Evaluation of high-temperature absorption coefficients of ionized gas plasmas in optical fibers, *IEEE Photon. Technol. Lett.* 22(3): 134–136.

Shuto, Y., Yanagi, S., Asakawa, S., Kobayashi, M. & Nagase, R. (2004a). Evaluation of high-temperature absorption coefficients of optical fibers, *IEEE Photon. Technol. Lett.* 16(4): 1008–1010.

Shuto, Y., Yanagi, S., Asakawa, S., Kobayashi, M. & Nagase, R. (2004b). Fiber fuse phenomenon in step-index single-mode optical fibers, *IEEE J. Quantum Electronics* 40(8): 1113–1121.

Takenaga, K., Omori, S., Goto, R., Tanigawa, S., Matsuo, S. & Himeno, K. (2008). Evaluation of high-power endurance of bend-insensitive fibers, *Proceedings of Optical Fiber Communication/National Fiber Optic Engineers Conference*. (JWA11).

Takenaga, K., Tanigawa, S., Matsuo, S., Fujimaki, M. & Tsuchiya, H. (2008). Fiber fuse phenomenon in hole-assisted fibers, *Proceedings of the 34th European Conference on Optical Communication*, Vol. 5, pp. 27–28. (P.1.14).

Todoroki, S. (2005a). Animation of fiber fuse damage, demonstrating periodic void formation, *Opt. Lett.* 30(19): 2551–2553.

Todoroki, S. (2005b). In-situ observation of fiber-fuse propagation, *Jpn. J. Appl. Phys.* 44(6A): 4022–4024.

Todoroki, S. (2005c). Origin of periodic void formation during fiber fuse, *Optics Express* 13(17): 6381–6389.

Todoroki, S. (2005d). Transient propagation mode of fiber fuse leaving no voids, *Optics Express* 13(23): 9248–9256.

Todoroki, S. (2008). In situ observation of modulated light emission of fiber fuse synchronized with void train over hetero-core splice point, *PLoS ONE* 3(9): e3276.

Todoroki, S. (2011). Threshold power reduction of fiber fuse propagation through a white tight-buffered single-mode optical fiber, IEICE Electronics Express 8 (accepted).

Wang, J., Gray, S., Walton, D. & Zenteno, L. (2008). Fiber fuse in high-power optical fiber, *in* M.-J. Li, P. Shum, I. H. White & X. Wu (eds), *Passive Components and Fiber-based Devices V*, Vol. 7134 of *SPIE Proceedings*, SPIE, pp. 71342E–1–9. (Hangzhou, China).

Yakovlenko, S. I. (2004). Plasma behind the front of a damage wave and the mechanism of laser-induced production of a chain of caverns in an optical fibre, *Quantum Electron.* 34(8): 765–770.

Yakovlenko, S. I. (2006a). Mechanism for the void formation in the bright spot of a fiber fuse, *Laser Physics* 16(3): 474–476.

Yakovlenko, S. I. (2006b). Physical processes upon the optical discharge propagation in optical fiber, *Laser Physics* 16(9): 1273–1290.

Yamada, M., Koyama, O., Katsuyama, Y. & Shibuya, T. (2011). Heating and burning of optical fiber by light scattered from bubble train formed by optical fiber fuse, *Proceedings of Optical Fiber Communication/National Fiber Optic Engineers Conference*. (JThA1).

Yanagi, S., Asakawa, S., Kobayashi, M., Shuto, Y. & Naruse, R. (2003). Fiber fuse terminator, *The 5th Pacific Rim Conference on Lasers and Electro-Optics*, Vol. 1, p. 386. (W4J-(8)-6, Taipei. Taiwan, 22-26 Jul. 2003).

6

Radiation Induced by Charged Particles in Optical Fibers

Xavier Artru and Cédric Ray

Université de Lyon, Université Lyon 1, CNRS-IN2P3,
Institut de Physique Nucléaire de Lyon
France

1. Introduction

The electric field of a charged particle passing through or near an optical fiber induces a transient charges and currents in the fibrer medium (1; 2). These charges and current radiates electromagnetic waves, both outside the fiber (free light) and inside (guided light). This chapter is devoted to the guided light, which will be referred to as PIGL, for *Particle Induced Guided Light*.

If the fiber radius is large enough and the particle passes trough it, as in Fig. 1, both PIGL and oustide radiation can be considered as transition radiation and becomes Cherenkov radiation when the particle velocity exceeds that of light in the medium. This is the basis of the quartz fibre particle detectors (3–5). Let us mention two other uses of optical fibers as particle detectors : (i) as dosimeters, through the effect of darkening by irradiation (6); (ii) in scintillating glass fibers for particle tracking.

Here we will consider fibers of radius a comparable to the wavelength, in which case the standard OTR or Cherenkov descriptions are not appropriate. Two types of PIGL have to be considered :

- Type I : The particle passes *near* or *through* a straight or weakly bent part of the fibre, far from an extremity. Translation invariance along the fiber axis is essential.
- Type II : The particle passes near or through an end of the fiber or an added structure (e.g., metallic balls glued on the fibre surface), which is not translation invariant.

2. Particle-induced guided light of Type-I

The PIGL intensity will be calculated in the framework of quantized fields used by Glauber (7). We will use relativistic quantum units units familiar to particle physicists : $\hbar = c = \varepsilon_0 = \mu_0 = 1$. $\lambdabar \equiv \lambda/2\pi = 1/\omega$. The Gauss law is written $\nabla \cdot \mathbf{E} = \rho$, not $4\pi\rho$. $e^2/(4\pi) = \alpha = 1/137$.

2.1 Expansion of the field in proper modes

The fiber is along the $\hat{z}$ axis. The cylindrical coordinates are (r, ϕ, z). $\mathbf{r} = (x, y)$ is the transverse position. $x \pm iy = re^{\pm i\phi}$.

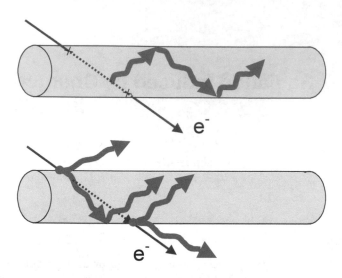

Fig. 1. Standard mechanisms of production of light inside a fiber, by an electron passing through. Top: Cherenkov radiation. Bottom: Transition radiation.

The quantized electromagnetic field $\mathbf{E}^{op}$ in presence of the fiber can be expanded in propagation modes :

$$\mathbf{E}^{op}(t, \mathbf{X}) = \int_0^\infty \frac{d\omega}{2\pi} \sum_m a_m(\omega)\, \vec{\mathcal{E}}^{(m)}(\omega; \mathbf{X})\, \exp(-i\omega t) + \text{hermit. conj.} \tag{1}$$

The complex-valued field

$$\vec{\mathcal{E}}^{(m)}(\omega; \mathbf{X}) = \mathbf{E}^{(m)}(\omega; \mathbf{r})\, \exp(ipz) \tag{2}$$

is a "photon wave function". $m = \{M, \nu, \sigma\}$ is a collective index which gathers the total angular momentum $M \equiv J_z = L_z + S_z$ of the photon, the radial quantum number ν and the direction of propagation $\sigma = \text{sign}(p) = \pm 1$. a_m and $a_m^\dagger$ are the destruction and creation operators of a photon in the mode m. ω and p are linked by the dispersion relation,

$$\omega = \omega_m(p) \quad \text{or} \quad p = p_m(\omega). \tag{3}$$

The ν spectrum has a discrete part for guided modes and a continuous part for free modes. The summation over m in (1) implies that ν is treated as a fully discrete variable, for simplicity. This is actually the case if we quantize the field inside a cylindrical box.
The quantized magnetic field is expanded like in (1). a_m and $a_m^\dagger$ obey the commutation rules

$$\left[a_{M,\nu,\sigma}(\omega), a_{M',\nu',\sigma'}^\dagger(\omega')\right] = 2\pi\, \delta(\omega - \omega')\, \delta_{MM'}\, \delta_{\nu\nu'}\, \delta_{\sigma\sigma'}. \tag{4}$$

For a fixed ω the modes m are orthonormal in the sense

$$\int d^2\mathbf{r}\, \left[\mathbf{E}^{(m)*}(\omega; \mathbf{r}) \times \mathbf{B}^{(n)}(\omega; \mathbf{r}) + \mathbf{E}^{(n)}(\omega; \mathbf{r}) \times \mathbf{B}^{(m)*}(\omega; \mathbf{r})\right]_z = \omega\, \delta_{mn}. \tag{5}$$

For $n = m$, the left-hand side is the power carried by the fiber in the mode m, which is $\hbar\omega$ (= one photon) per unit of time.

Equations (1), (4) and (5) correspond to Eqs. (2.29b), (2.25b) and (2.14a) of Ref.(7). The correspondance would be $f_{\mathbf{k}} \rightarrow -i\,(2/\omega)^{1/2}\,\mathbf{E}^{(m)}$, but we use the continuous variable ω instead of a fully discrete set of quantum numbers. a_m and $\mathbf{E}^{(m)}$ differ from those of Ref.(1) by a factor $(dp/d\omega)^{1/2} = v_g^{-1/2}$. The factor 2 in (5) was forgotten in Refs.(1; 2), leading to an overestimation of the photon production yield by a factor 2.

2.2 Wave functions of the fiber modes

The propagation modes in optical fibers can be found in several textbooks, e.g. (8). Nevertheless, it is useful to present a short review based on states of definite angular momentum M.

We assume that the fiber has an homogeneous refractive index $n = \sqrt{\varepsilon}$ and no clad. For a guided mode the phase velocity $v_{\mathrm{ph}} = \omega/p$ is in the interval $[1/n, 1]$. The photon transverse momentum is $q = \sqrt{\varepsilon\omega^2 - p^2}$ inside the fiber and $i\kappa = i\sqrt{p^2 - \omega^2}$ (evanescent wave) outside the fiber. The longitudinal parts of the fields have $S_z = 0$ therefore their orbital angular momentum L_z is equal to M. Using cylindrical coordinates (r, ϕ, z) they write

$$E_z(\mathbf{r}) = i\,e^{iM\phi}\,f_z(r), \qquad B_z(\mathbf{r}) = e^{iM\phi}\,h_z(r), \tag{6}$$

Both in medium and in vacuum f_z and h_z obey the same differential equation

$$\left[\partial_r^2 + r^{-1}\partial_r - M^2/r^2 + k_T^2(r)\right] f_z \text{ or } h_z = 0 \quad (\text{except for } r = a) \tag{7}$$

where $k_T^2(r) = q^2$ inside the fiber and $k_T^2(r) = -\kappa^2$ outside the fiber.

The piecewise solutions of (7) are Bessel functions J_M or K_M. From the fact that f_z and h_z are continuous at $r = 0$ and $r = a$ and decreasing at $r \rightarrow \infty$, it follows that $h_z(r)/f_z(r)$ is independent on r. We write

$$f_z(r) = c_E\,\psi(r), \qquad h_z(r) = c_B\,\psi(r), \tag{8}$$

$$\psi(r) = J_M(qr) \text{ inside}, \quad \psi(r) = c_K\,K_M(\kappa r) \text{ outside}, \quad c_K = \frac{J_M(qa)}{K_M(\kappa a)}.$$

The transverse components $\mathbf{E}_T$ and $\mathbf{B}_T$ can be expressed either in terms of the radial and azimuthal basic vectors, $\hat{\mathbf{e}}^r = \mathbf{r}/r$ and $\hat{\mathbf{e}}^\phi = \hat{\mathbf{z}} \times \hat{\mathbf{e}}^r$,

$$\mathbf{E}_T = e^{iM\phi}\,\left(f_r(r)\,\hat{\mathbf{e}}^r + f_\phi(r)\,\hat{\mathbf{e}}^\phi\right)$$

$$\mathbf{B}_T = e^{iM\phi}\,\left(h_r(r)\,\hat{\mathbf{e}}^r + h_\phi(r)\,\hat{\mathbf{e}}^\phi\right), \tag{9}$$

or in terms of the $S_z = \pm 1$ eigenvectors $\hat{\mathbf{e}}^\pm = (\hat{\mathbf{x}} \pm i\hat{\mathbf{y}})/2$:

$$\mathbf{E}_T = e^{i(M-1)\phi}\,f_-(r)\,\hat{\mathbf{e}}^+ + e^{i(M+1)\phi}\,f_+(r)\,\hat{\mathbf{e}}^-$$

$$i\,\mathbf{B}_T = e^{i(M-1)\phi}\,h_-(r)\,\hat{\mathbf{e}}^+ + e^{i(M+1)\phi}\,h_+(r)\,\hat{\mathbf{e}}^- \tag{10}$$

with $f_\pm = f_r \pm i f_\phi$ and $-i h_\pm = h_r \pm i h_\phi$. The $\hat{e}^+$ and $\hat{e}^-$ parts of the fields have orbital momenta $L_z = M \mp 1$, therefore their radial dependence are Bessel functions of order $M \mp 1$:

$$f_\pm(r) = c_{fJ}^\pm J_{M\pm1}(qr)\ (r \le a), \qquad c_{fK}^\pm K_{M\pm1}(\kappa r)\ (r > a),$$
$$h_\pm(r) = c_{hJ}^\pm J_{M\pm1}(qr)\ (r \le a), \qquad c_{hK}^\pm K_{M\pm1}(\kappa r)\ (r > a). \tag{11}$$

The Maxwell equations relate the transverse fields to the longitudinal ones. The formula in the $\{\hat{e}^r, \hat{e}^\phi\}$ basis can be found in (8). Translated in the $\{\hat{e}^+, \hat{e}^-\}$ basis they give

$$c_{fJ}^\pm = (\pm p\, c_E - \omega\, c_B)/q, \qquad c_{fK}^\pm = \mp(q\, c_K/\kappa)\, c_{fJ}^\pm,$$

$$c_{hJ}^\pm = (\pm p\, c_B - \omega \varepsilon\, c_E)/q, \qquad c_{hK}^\pm = (-p\, c_B \pm \omega\, c_E)\, c_K/\kappa.$$

The continuity of h_z, h_r, h_ϕ, f_z, f_ϕ and $\varepsilon(r) f_r$ at $r = a$ leads to

$$\frac{c_B}{c_E} = -MQ\left[\frac{J'_M(u)}{u J_M(u)} + \frac{K'_M(w)}{w K_M(w)}\right]^{-1} = -\frac{1}{MQ}\left[\frac{\varepsilon J'_M(u)}{u J_M(u)} + \frac{K'_M(w)}{w K_M(w)}\right] \tag{12}$$

where $u \equiv qa$, $w \equiv \kappa a$ and

$$Q = \left(u^{-2} + w^{-2}\right) p/\omega = \left(\varepsilon u^{-2} + w^{-2}\right)\omega/p.$$

From the two expressions of c_B/c_E in (12) one obtains

$$\left[\frac{J'_M(u)}{u J_M(u)} + \frac{K'_M(w)}{w K_M(w)}\right] \cdot \left[\frac{\varepsilon J'_M(u)}{u J_M(u)} + \frac{K'_M(w)}{w K_M(w)}\right] = M^2 \left(\frac{1}{u^2} + \frac{1}{w^2}\right) \cdot \left(\frac{\varepsilon}{u^2} + \frac{1}{w^2}\right), \tag{13}$$

which, together with $u^2 = (\varepsilon\omega^2 - p^2)a^2$ and $w^2 = (p^2 - \omega^2)a^2$, determines the dispersion relation (3).

2.2.1 Normalization of the mode wave functions

The z-component of the Pointing vector of the complex field is

$$\mathcal{P}^{(m)}(\mathbf{r}) = 2\,\mathrm{Re}\left\{\mathbf{E}^{(m)*} \times \mathbf{B}(m)\right\}_z = \mathrm{Re}\left\{f_-^*(r)\, h_-(r) - f_+^*(r) h_+(r)\right\}.$$

Using (11) and integrating over $\mathbf{r}$ gives the mode power

$$P^{(m)} = P_{\mathrm{int}}^{(m)} + P_{\mathrm{ext}}^{(m)} = \int_0^a 2\pi r\, dr \left\{c_{fJ}^- c_{hJ}^-\, J_{M-1}^2(qr) - c_{fJ}^+ c_{hJ}^+\, J_{M+1}^2(qr)\right\}$$
$$+ \int_a^\infty 2\pi r\, dr \left\{c_{fK}^- c_{hK}^-\, K_{M-1}^2(\kappa r) - c_{fK}^+ c_{hK}^+\, K_{M+1}^2(\kappa r)\right\}. \tag{14}$$

The coefficient c_E has to be adjusted to get the normalization (5).

Fig. 2 shows the phase velocity $v_{\mathrm{ph}} = \omega/p$ of the lowest mode ($M = \pm 1, \nu = 1$) called HE_{11} and the external fraction of the mode power, as a function of ω. The index of refraction is $n = 1.41$ (fused silica).

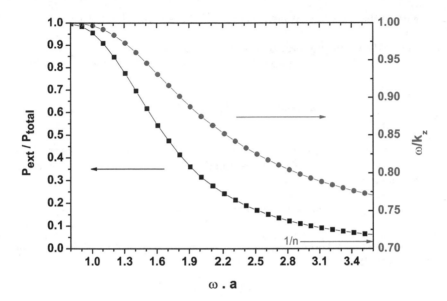

Fig. 2. Phase velocity $v_{ph} = \omega/p$ (balls, right scale) and external fraction of the power (squares, left scale) for the HE_{11} mode.

2.2.2 Linearly polarized modes

When changing M into $-M$, the above defined field modes change as follows :

$$\{E_T, E_z, \mathbf{B}_T, B_z\}^{(-M)} = (-1)^M \; \Pi(0°) \; \{E_T, E_z, \mathbf{B}_T, B_z\}^{(M)}$$
$$= \Pi(90°) \; \{E_T, E_z, \mathbf{B}_T, B_z\}^{(M)}$$
$$= (-1)^M \; \{\mathbf{E}_T^*, -E_z^*, \mathbf{B}_T^*, -B_z^*\}^{(M)}. \qquad (15)$$

$\Pi(\alpha)$ is the operator of mirror reflection about the plane $\phi = \alpha$, for instance

$$\Pi(0°)\{E_x, E_y, E_z\}(x, y, z) = \{E_x, -E_y, E_z\}(x, -y, z)$$

and a similar formula for $\mathbf{B}$, with an extra $(-)$ sign since it is a pseudovector. The linear combination

$$\{\mathbf{E}, \mathbf{B}\}^{(M,0°)} = \left[\{\mathbf{E}, \mathbf{B}\}^{(M)} + (-1)^M \{\mathbf{E}, \mathbf{B}\}^{(-M)}\right]/\sqrt{2} \qquad (16)$$

is even under $\Pi(0°)$ and has real $\mathbf{E}_T$. For $M = 1$,

$$\mathbf{E}_T^{(1,0°)} = [f_-(r)\,\hat{\mathbf{x}} + f_+(r)\,(\cos 2\phi\,\hat{\mathbf{x}} + \sin 2\phi\,\hat{\mathbf{y}})]/\sqrt{2} \qquad (17)$$

is the state whose dominant (f_-) part is linearly polarized parallel to $\hat{\mathbf{x}}$.

2.3 Bent fiber

Bending the fiber has several effects :
- a) small break-down of the degeneracy (*i.e.*, slightly different dispersion relations) between the polarized states $(M, 0°)$ and $(M, 90°)$, where $0°$ is the azimuth of the bending plane,

- b) co-rotation of the transverse wave function $\vec{\mathcal{E}}^{(m)}(\omega; \mathbf{X})$ with the unit vector $\hat{s}$ tangent to the local fiber axis.
- c) escape of light by tunneling through a centrifugal barrier.
For large enough bending radius, effects a) and c) can be ignored. Effect b) is non-trivial when the bending is skew (not planar). Instead of (2), we have

$$\vec{\mathcal{E}}^{(m)}(\omega; \mathbf{X}) = \mathcal{R}_f(s)\, \mathbf{E}^{(m)}\left(\omega; \mathcal{R}_f^{-1}(s)\, \mathbf{r}\right)\, \exp(ips),\tag{18}$$

where $\mathbf{X}_f(s)$ is the point of the fiber axis nearest to $\mathbf{X}$, s its curvilinear abscissa and $\mathbf{r} = \mathbf{X} - \mathbf{X}_f(s)$ (see Fig. 3 left). $\mathcal{R}_f(s)$ is a finite rotation matrix resulting from a succession of infinitesimal rotations $\mathcal{R}(\hat{s} \to \hat{s} + d\hat{s})$:

$$\mathcal{R}_f(s + ds) = \mathcal{R}(\hat{s} \to \hat{s} + d\hat{s}) \circ \mathcal{R}_f(s), \quad \mathcal{R}_f(0) = I,\tag{19}$$

$\mathcal{R}(\hat{s} \to \hat{s}')$ denoting the rotation along $\hat{s} \times \hat{s}'$ which transforms $\hat{s}$ into $\hat{s}'$. Taking into account the non-commutativity of the rotations, we have

$$\mathcal{R}_f(s) = \mathcal{R}(\hat{s}, \Omega(s)) \circ \mathcal{R}(\hat{z} \to \hat{s}).\tag{20}$$

where $\hat{z}$ is the orientation of the beginning of the fiber, $\mathcal{R}(\hat{s}, \alpha)$ stands for a rotation of angle α about $\hat{s}$ and $\Omega(s)$ is the dark area on the unit sphere in Fig. 3 (right). For a state of given angular momentum M in (18) one can replace $\mathcal{R}_f(s)$ by $\mathcal{R}(\hat{z} \to \hat{s})$ and take into account the first factor of (20) by the *Berry phase* factor $\exp[-iM\Omega(s)]$. If the fiber is bent in a plane, $\Omega(s) = 0$.

2.4 Mode excitation by a charged particle
When a particle of charge Ze passes *trough* or *near* the fiber, it can create one or several photons by spontaneous or stimulated emission. Neglecting its loss of energy and momentum, the particle acts like a cassical current and the excitation of the quantum field is a *coherent state* (7). The spontaneous photon emission amplitude in the mode m, corresponding to Eqs. (7.11) and (7.16) of (7), is

$$R^{(m)}(\omega) = \frac{Ze}{\omega} \int d\mathbf{X}(t) \cdot \vec{\mathcal{E}}^{(m)*}(\omega; \mathbf{X})\, \exp(i\omega t)\tag{21}$$

for a mode normalized according to (5). The photon spectrum of spontaneous emission in the mode m reads

$$\frac{d\mathcal{N}_{\text{phot}}^{(m)}}{d\omega} = \frac{\omega}{2\pi P^{(m)}(\omega)} \left|R^{(m)}(\omega)\right|^2.\tag{22}$$

Thanks to the factor $P^{(m)}(\omega)$ given by (14) in the denominator, this expression is invariant under a change of the normalisation of the mode fields.

2.5 Straight fiber and particle in rectilinear uniform motion
For a particle following the straight trajectory

$$\mathbf{X} = \mathbf{b} + \mathbf{v}t, \quad \mathbf{b} = (b, 0, 0), \quad \mathbf{v} = (0, v_T, v_L),\tag{23}$$

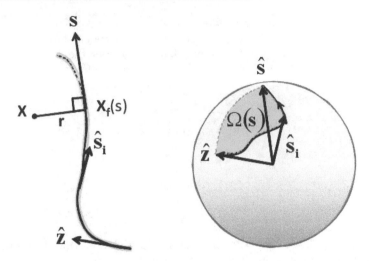

Fig. 3. Left: bent fiber and definition of $\mathbf{X}_f(s)$ and $\mathbf{r}$. Right: curve drawn by the extremities of successive tangent vectors ($\hat{\mathbf{z}}$, $\hat{\mathbf{s}}_i$, $\hat{\mathbf{s}}$, ...) on the unit sphere and definition of the solid angle $\Omega(s)$. The dotted arc of circle represents the "most direct" rotation, $\mathcal{R}(\hat{\mathbf{z}} \rightarrow \hat{\mathbf{s}})$, transforming $\hat{\mathbf{z}}$ int $\hat{\mathbf{s}}$.

Eqs.(21) and (2) give

$$R^{(m)}(\omega) = \frac{Ze}{\omega} \int_{-\infty}^{\infty} dy \left[E_y^{(m)}(x,y) + \frac{v_L}{v_T} E_z^{(m)}(x,y) \right]^* \exp\left(iy \frac{\omega - v_L p}{v_T} \right) \tag{24}$$

Using (6-11) one arrives at the pure imaginary expression

$$R^{(m)}(\omega) = \frac{-iZe}{\omega} \int_0^{\infty} dy \left\{ \cos[\eta y + (M-1)\phi] \, f_-(r) \right.$$
$$\left. - \cos[\eta y + (M+1)\phi] \, f_+(r) + 2(v_L/v_T) \cos(\eta y + M\phi) \, f_z(r) \right\} \tag{25}$$

with $r = \sqrt{b^2 + y^2}$, $\phi = \tan^{-1}(y/b)$ and

$$\eta = (v_L \, p - \omega)/v_T = (v_L - v_{ph}) \, p/v_T .$$

2.6 Limit of small crossing angle

For small crossing angle $\theta = \tan^{-1}(v_T/v_L)$ the integrand of (24) becomes large due to the v_L/v_T factor of the third term, although f_z is generally small. On the other hand, unless $|v_L - v_{ph}| \lesssim v_T/(pa)$, the integrand oscillates fast in the region $|y| \lesssim a$ where the field is important and the amplitude is strongly reduced. One therefore expects an almost monochromatic peak at $\omega = \omega_C(v)$ fixed by the "fiber Cherenkov condition"

$$v_{ph}(\omega_C) \equiv \omega_C/p(\omega_C) = v \tag{26}$$

and the dispersion relation (3). The case $\theta = 0$ (electron runnig parallel to the fiber), where $\omega \equiv \omega_C$, has been studied in Refs.(9; 10).

2.7 Slightly bent fiber or particle trajectory

Local curvatures of the trajectory or of the fiber can be neglected and formula (25) is accurate enough when the crossing angle θ is large. Let us consider the case where the particle trajectory, the fiber or both are slightly curved, but at angles not far from the $\hat{z}$ direction. Then we have to use (18) instead of (2) in (21). However we can omit the rotation matrix $\mathcal{R}_f(s)$ and make the approximation

$$d\mathbf{X}(t) \cdot \vec{\mathcal{E}}^{(m)*}(\omega; \mathbf{X}) \simeq v\, dt\, E_z^{(m)}(\omega; \mathbf{r})\, \exp(ips). \tag{27}$$

Thus we can rewrite (21) as

$$R^m(\omega) = \frac{Zev}{\omega} \int dt\, E_z^*[(\omega; \mathbf{r}(t)]\, \exp[i\omega t - ip_m(\omega)s],$$

$$\mathbf{r}(t) = \mathbf{X}_p(t) - \mathbf{X}_f(s), \quad s = \int v\, dt\, \cos\theta(t). \tag{28}$$

Here again the integrand oscillates too fast - and the amplitude is too small - when ω is not close to ω_C. The total photon number in the mode m is

$$\mathcal{N}_{phot}^{(m)} = 2Z^2\, \alpha\, v^2 \int \frac{d\omega}{\omega\, P^{(m)}(\omega)} \int dt'\, E_z^*[(\omega; \mathbf{r}(t')] \int dt''\, E_z[(\omega; \mathbf{r}(t'')] \tag{29}$$

$$\exp\{i\omega(t' - t'') - ip(\omega)\,(s' - s'')\}. \tag{30}$$

To first order in $\omega - \omega_C$ the exponential can be written as

$$\exp\{i\omega\,(T - S/v_g) + i[\omega_C/v_g - p_m(\omega_C)]\,S\} \tag{31}$$

where $t' - t'' = T$, $s' - s'' = S$ and $v_g = d\omega/dp$ is the group velocity at $\omega = \omega_C$. Neglecting the variations of the other factors with ω, the integration over ω yields a factor $2\pi\delta(T - S/v_g)$. From the second line of (28), we have $S/T \simeq v\cos\theta(t) \simeq v$ at small S and T, therefore $\delta(T - S/v_g) = \delta(T)/[1 - v/v_g]$. One finally obtains

$$\mathcal{N}_{phot}^{(m)} = \frac{4\pi Z^2\alpha\, v^2}{\omega_C\, P^{(m)}(\omega_C)} \frac{1}{|1 - v/v_g(\omega_C)|} \int dt\, |E_z[\omega_C; \mathbf{r}(t)]|^2. \tag{32}$$

The energy of the light pulse is obtained by multiplying by ω_C. This formula applies in particular to the limit of small crossing angles considered above. The photon number increases linearly with the path length over which the particle travels inside or close to the fiber.

2.8 Numerical results for straight electron trajectory and straight fiber

The dimensionless photon spectrum $\omega d\mathcal{N}_{phot}/d\omega$ in the fundamental mode HE_{11} of a fused silica fiber is plotted in Fig. 4 for three impact parameters, $b = 0.2\,a$ (penetrating trajectory), $b = a$ (tangent trajectory) and $b = 1.5\,a$ (fully external trajectory), and two particle velocity vectors, $(v_L, v_T) = (0.88, 0.1)$ and $(v_L, v_T) = (0.85, 0.5)$, corresponding to large and moderate angle respectively. We took the sign of M to be the same as the J_z of the particle.

The spectra are harder for penetrating trajectories, due to (i) the discontinuity of the fields at the fiber surface, (ii) the lower importance of the evanescent field at high frequency.

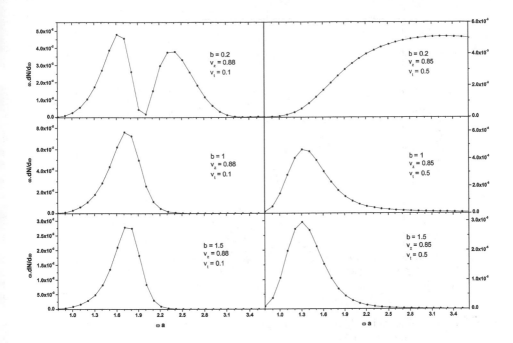

Fig. 4. Dimensionless photon spectrum $\omega d\mathcal{N}_{\text{phot}}/d\omega$ as a function of ωa in the HE_{11} mode for six types of the particle trajectory and $M = +1 = \text{sign}(v_T)$.

In the large angle - penetrating case, the dimensionless yield is of the order of $\alpha = e^2/(4\pi) = 1/137$. In the tangent case it is much smaller. Note the peak at a relatively small frequency, where the wave travels mainly outside the fiber (see Fig. 2). At still smaller frequency, the wave function of the mode becomes too much diluted, which explains the vanishing yields at small ω in the six curves.

In the $b = 0.2\,a$ and $v_T = 0.1$ case, we have a dip at $\omega a = 2$ instead of an expected Cherenkov peak fixed by Eq.(26). This is a peculiarity of the odd M modes when b is small : if $b = 0$, then ϕ in (25) is either $-\pi/2$ or $+\pi/2$ and, at the Cherenkov point ($\eta = 0$), $\cos(\eta y + M\phi)$ is zero in the whole integration range.

A separate figure (Fig. 5) at small crossing angle ($v_T/v_L = 0.03/0.95$) shows the narrow peak of "fiber Cherenkov light" at the position $\omega a \simeq 1.4$ predicted by (26) and Fig. 2. The half-width at half maximum, 0.06, corresponds roughly to the condition $|v_L - v_{\text{ph}}| \lesssim v_T/(pa)$ mentioned in Paragraph 2.6.

2.9 Polarisation

If $b = 0$, the HE_{11} guided light is linearly polarized in the particle incidence plane. If $b \neq 0$, some circular polarization is expected. One could naively expect that the favored photon angular momentum M has the sign of the azimutal speed of the particle, i.e. the sign of v_T in (23), but this is not always true. What matters in fact is not the sign of M but the sense of rotation of the electric field of the mode in the *moving plane* $z = vt$. In this plane the azimuth of the field varies like $M(\omega t - pz) = (v_{\text{ph}} - v_L) M\omega t/v_{\text{ph}}$. If the moving plane is faster than

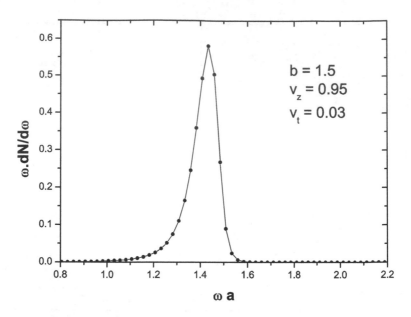

Fig. 5. Photon yields in the HE_{11} mode with $M = +1$ for a small crossing angle :
$(v_L, v_T) = (0.95, 0.03)$; $b = 1.5a$.

the wave, the field rotates in the opposite way. Thus the favored sign of M is the sign of $(v_{ph} - v_L) v_T$. This can be seen from (25) : if M and η have the same sign, the integrand oscillate faster and the amplitude is reduced.

In Figs. 4 and 5, M has the sign of v_T. This circular polarization is favored at $v_{ph} > v_L$, whence $\omega < \omega_C(v_L)$, and unfavored at $v_{ph} < v_L$, whence $\omega > \omega_C(v_L)$. This partly explains the asymmetric shape of the fiber Cherenkov peak in Fig. 5. Changing the sign either of M or of v_T should result in a harder spectrum.

2.10 Interferences with periodically bent trajectory or bent fiber

With an undulated trajectory, as in Fig. 6a or an undulated fiber as in Fig. 6b, one can have several meeting points, the PIGL amplitude of which, given by (25) or (28), add coherently. Let L_f and L_p be the lengths of the fiber and of the particle trajectory between two meeting points. Two successive fiber-particle interactions are separated in time by $\Delta t = L_f/v$ and their phase difference is

$$\Delta\Phi = p\, L_f - \omega\, \Delta t = \omega\, \left(L_f/v_{ph} - L_p/v\right).\tag{33}$$

If N equivalent meeting points are spaced periodically, the frequency spectrum is

$$\left(\frac{d\mathcal{N}^{(m)}}{d\omega}\right)_{N\text{ meeting}} = \left(\frac{d\mathcal{N}^{(m)}}{d\omega}\right)_{1\text{ meeting}} \times \frac{\sin^2(N\Delta\Phi/2)}{\sin^2(\Delta\Phi/2)}.\tag{34}$$

The last fraction is the usual interference factor in periodical systems, e.g. in undulator radiation. For large N it gathers the photon spectrum in quasi-monochromatic lines fixed by

$$\omega\left(L_f/v_{\mathrm{ph}} - L_p/v\right) = 2k\pi \quad (k\text{ integer})\,. \tag{35}$$

If the fiber bending is *not planar*, but for instance helicoidal (Fig. 6c), the left- and right circular polarisations have different phase velocities. Their propagation amplitudes acquire an additional phase $\phi_B = -M\Omega$, called the *Berry phase*, where Ω is the solid angle of the cone drawn by the local axis of the fiber (11) (as if $\hat{s}$ coincides with $\hat{z}$ in Fig. 3). The preceding condition becomes

$$\omega\left(L_f/v_{\mathrm{ph}} - L_p/v\right) = 2k\pi - \phi_B\,. \tag{36}$$

The interferences disappear when the velocity spread of the charged particle beam is such that the variation of $\omega\,L_p/v$ is more than, say, 2π.

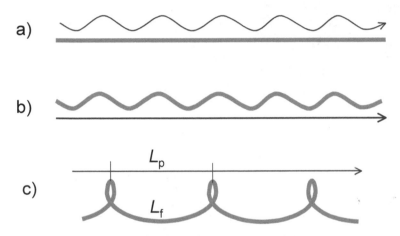

Fig. 6. Periodically bent particle trajectory (a), planar bent fiber (b) and helical bent fiber (c). L_p and L_f are the lengths of the curved or straight periods, for the particle and the fiber respectively.

2.11 Application of type-I PIGL to beam diagnostics

PIGL in a monomode fiber is intense enough not for single particle detection, but for beam diagnostics.

The "fiber Cherenkov radiation" can be used to measure the velocity of a semi-relativistic particle beam, using the dependence of v_{ph} on ω shown in Fig. 2.

In a periodically bent fiber, the interference can test the velocity spread of the beam.

At large crossing angle, a fiber can measure the transverse profile of the beam with a resolution of the order of the diameter $2a$. No background is made by real photons coming from distant sources (for instance synchrotron radiation from upstream bending magnets). Indeed, such photons are in the continuum spectrum of the radial number ν, therefore they are not captured by the fiber, but only scattered. This is an advantage over beam diagnostic tools like optical transition radiation (OTR) and optical diffraction radiation (ODR). The translation invariance along the fiber axis, which guarantees the conservation of ν, is essential for this property.

The resolution power of PIGL is also not degraded by the large transverse size $\sim \gamma \lambdabar$ of the virtual photon cloud at high Lorentz factor $\gamma = (1 - v^2)^{-1/2}$. Indeed, the virtual photons at transverse distance $\gg \lambdabar$ are almost real, therefore are not captured by the fiber.

3. Particle-induced guided light of Type-II

The second type of PIGL is produced at a place where the fiber is not translation invariant. We consider two examples : 1) PIGL from the cross section of a cut fiber, 2) PIGL assisted by metallic balls glued to the fiber. These devices are represented in Fig. 7.

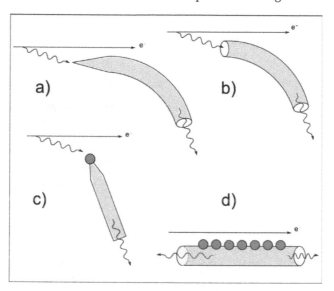

Fig. 7. Part of fiber which can capture virtual photons for Type-II PIGL : a) conical end ; b) sharp-cut end ; c) metallic ball glued on one end ; d) regularly spaced metallic balls glued along the fiber.

3.1 PIGL from the cross section of a cut fiber

The entrance section of a sharp-cut fiber can catch free real photons and convert them into guided photons. Assuming that the photons are incident at small angle with the fiber axis, the energy spectrum captured by the fiber in the mode $m = \{M, \nu\}$ is given by

$$\frac{dW^{(m)}}{d\omega} = \frac{1}{2\pi P^{(m)}(\omega)} \times \left| \int d^2\mathbf{r} \left[T_B(\mathbf{r}) \, \mathbf{E}_T^{(m)*}(\omega; \mathbf{r}) \times \mathbf{B}_T^{in}(\omega; \mathbf{r}) \right.\right.$$
$$\left.\left. + T_E(\mathbf{r}) \, \mathbf{E}_T^{in}(\omega; \mathbf{r}) \times \mathbf{B}_T^{(m)*}(\omega; \mathbf{r}) \right] \right|^2 . \tag{37}$$

where $\{\mathbf{E}^{in}, \mathbf{B}^{in}\}$ is the incoming field on the cutting plane. $T_E(\mathbf{r})$, $T_B(\mathbf{r})$ are the Fresnel refraction coefficients at normal incidence, given by

$$T_E(\mathbf{r}) = 2/(1 + \sqrt{\varepsilon(\mathbf{r})}), \quad T_B(\mathbf{r}) = \sqrt{\varepsilon(\mathbf{r})} \, T_E(\mathbf{r}). \tag{38}$$

$\varepsilon(\mathbf{r})$ is the local permittivity of the fiber. Outside the fiber, $T_E(\mathbf{r}) = T_B(\mathbf{r}) = 1$. Equation (37) is deduced from the orthonormalization relation (5).

With some caution (37) can be applied to the capture of virtual photons from the Coulomb field of a relativistic particle passing near the entrance face (see Fig. 7b). The transverse component of this field is given by (12; 13)

$$\mathbf{E}_T^{\text{in}}(\omega;\mathbf{r}) = \frac{Ze\,\omega}{2\pi\gamma v^2 b} K_1\left(\frac{\omega b}{\gamma v}\right)\mathbf{b}, \quad \mathbf{B}_T^{\text{in}}(\omega;\mathbf{r}) = \mathbf{v} \times \mathbf{E}_T^{\text{in}}(\omega;\mathbf{r}). \tag{39}$$

Here $\mathbf{b} = \mathbf{r} - \mathbf{r}_{\text{particle}}$ is the impact parameter relative to the particle. It must be large enough compared to $\lambdabar$, otherwise the incoming photon is too different from a real one.

3.2 PIGL from a conical end of fiber

The sharp-cut fiber has a wide angular acceptance but is not optimized for capturing the virtual photon cloud accompagning an ultrarelativistic particle, which has an angular divergence $\sim 1/\gamma$. A more efficient capture is possible with a narrow conical end (Fig. 7a), at the price of a smaller acceptance. The wave function of a parallel photon may be quasi-adiabatically transformed into a guided mode without too much loss. This should be true for the photons of the Coulomb field in the impact parameter range $\lambdabar \ll b \lesssim \gamma\lambdabar$, which are quasi-real and have a small transverse momentum $k_T \sim 1/b$.

3.3 PIGL from metallic balls

It is also possible to capture a virtual photon with a metallic ball glued to the fiber, either at the extremity (Fig. 7c) (14; 15), or on the side as in Fig. 7d. Then a plasmon is created (16; 18), which has some probability p_f to be evacuated as guided light in the fiber.

A rough estimate of the capture efficiency can be obtained when the impact parameter of the particle is large compared to the ball radius R and the time scale $\Delta t \sim b/(\gamma v)$ of the transient field is short compared to the reduced period $1/\omega = \lambdabar$ of the plasmon : the particle field boosts each electron of the ball with a momentum $\mathbf{q} \simeq 2Z\alpha\,\mathbf{b}/(vb^2)$. It results in a collective dipole excitation of the electron cloud, of energy

$$W(b) \simeq \frac{4\pi R^3 n_e}{3}\left(\frac{2Z\alpha}{vb}\right)^2 \frac{1}{2m_e} = \frac{2Z^2\alpha}{3v^2}\frac{\omega_p^2 R^3}{b^2} \quad (R \ll b \ll \gamma v\lambdabar), \tag{40}$$

where $\omega_p = (4\pi\alpha\,n_e/m_e)^{1/2}$ is the plasma frequency of the infinite medium. For a spherical ball the dipole plasmon frequency is simply given by $\omega = \omega_p/\sqrt{3}$, assuming the Drude formula $\varepsilon = 1 - \omega_p^2/\omega^2$ and neglecting the retardation effects (case $R \lesssim \lambdabar$). The number of stored quanta is then

$$\mathcal{N}(b) = \frac{W(b)}{\omega} \simeq \frac{2Z^2\alpha}{v^2}\cdot\frac{R^3}{\lambdabar\,b^2}. \tag{41}$$

Taking $b_{min} = R$ and $b_{max} = \gamma v\lambdabar$, the cross section for this process is

$$\sigma = \int_{b_{min}}^{b_{max}} 2\pi\,b\,db\,\mathcal{N}(b) \simeq \frac{4Z^2\alpha}{v^2}\cdot\frac{R^3}{\lambdabar}\cdot\ln\frac{\gamma v\lambdabar}{R}. \tag{42}$$

More precise values of the plasmon frequencies are used in (16–18) in the context of Smith-Purcell radiation. Retardation effects and other mutipoles are taken into account in (17; 18). A typical order of the cross section, $\sigma \sim 10^{-2} \lambdabar^2$ is obtained with $R \sim \lambdabar$, $Z = 1$, $\gamma v \sim 1$. The plasmon wavelength is typically $\lambdabar \sim 10^2$ nm. Larger cross section can be realized by increasing R, but higher multipoles will dominate, unless γ is increased simultaneously. Discussions and experimental results about this point are given in (18).

The efficiency of the ball scheme depends on the ball-to-fiber transmission probability p_f, which is less than unity because the plasmon may also be radiated in vacuum or decay by absorption in the metal.

3.3.1 Interferences between several balls

If several metallic balls are glued at equal spacing l on one side of the fiber (Fig. 7d), constructive interferences (resonance peaks) are obtained when

$$\omega/v \mp p \equiv (1/v \mp 1/v_{\mathrm{ph}})\omega = 2k\pi/l \quad (k \text{ integer}), \tag{43}$$

ω and p being linked by (3). The $-$ and $+$ signs correspond respectively to lights propagating forward and backward in the fiber. The forward light has the highest frequency. This process is in competition with the Smith-Purcell radiation from the balls, where $\mp 1/v_{\mathrm{ph}}$ is replaced by $-\cos\theta_{\mathrm{rad}}$. We can call it "guided Smith-Purcell" radiation. It is advantageous to choose l such that ω lies on a plasmon resonance of the ball.

3.3.2 Shadowing

The guided Smith-Purcell spectrum for N balls can be written as

$$\left(\frac{d\mathcal{N}^{(m)}}{d\omega}\right)_{N\,\mathrm{balls}} \simeq \left(\frac{d\mathcal{N}^{(m)}}{d\omega}\right)_{1\,\mathrm{ball}} \times \frac{\sin^2(N\Delta\Phi/2)}{\sin^2(\Delta\Phi/2)} \times \text{shadow factor}. \tag{44}$$

This is similar to (34) except for a *shadow factor* which is less than unity. Indeed, each ball intercepts part of the virtual photon flux, thus makes a shadow on the following balls. The shadow of one ball has a longitudinal extension $l_f \sim v\lambda/(1-v) \sim \gamma^2 v\lambda$. Beyond this region, called *formation zone*, the cloud of virtual photons of wavelength λ is practically restored if there is no other piece of matter in the formation zone.

The shadow effect has been directly observed in diffraction radiation (19). In the case of mettalic balls it is included in the rescattering effects studied by García et al (20).

3.4 Application of Type-II PIGL to beam diagnostics

Type-II PIGL captures real as well as virtual photons : it acts both as a *near field* and a *far field* detector. Type-II PIGL can therefore be used for beam monitoring, but, like OTR and ODR, it is sensitive to backgrounds from distant radiation sources.

If the particle beam is ultrarelativistic, the quasi-real photons of the Coulomb field at impact parameter up to $b_{max} \sim \gamma\lambdabar$ can be captured. They give the logarithmic increase of (42) with γ and a similar one in (37). They can degrade somewhat the resolution power of Type-II PIGL in transverse beam size measurements, but experience with OTR monitors shows that this effect is not drastic (21–24).

4. Conclusion

This chapter shows the various possibilities of optical fibers in charged particle beam physics. The phenomenon of light production by a particle passing *near* the fiber, which has some theoretical interest, has not been tested experimentally up to now.

The flexibility of a fiber is an advantage over the delicate optics of OTR and ODR. A narrow fiber has less effects on the beam emittance than the metallic targets used in OTR and ODR.

Much work remains to be done before using the Type-I and Type-II PIGL : find the most convenient wavelength domain (infra-red, visible or ultraviolet) and fiber diameter ; determine the ball-to-fiber transmission coefficients p_f, etc.

The fiber has to be monomode if one wants to emphasize the interference effects. However it would be interesting to make simulations and experiments of the excitations of modes higher than HE_{11}. In particular the $M = 0$ TM mode has a significant E_z component, therefore may be excited at small crossing angle as much as the HE_{11} mode.

5. References

[1] X. Artru and C. Ray, Ed. S.B. Dabagov, Proc. SPIE, Vol. 6634 (2007) ; arXiv: hep-ph/0610129.

[2] X. Artru and C. Ray, Nucl. Inst. Meth. in Phys. Research B 266 (2008) 3725.

[3] P. Coyle et al, Nucl. Instr. Methods in Phys. Research A 343 (1994) 292.

[4] A. Contin, R. De Salvo, P. Gorodetzky, J.M. Helleboid, K.F. Johnson, P. Juillot, D. Lazic, M. Lundin, Nucl. Instr. Methods in Phys. Research A 367 (1994) 271.

[5] E. Janata, Nucl. Inst. Meth. in Phys. Research A493 (2002) 1.

[6] H. Henschel, M. Körfer, J. Kuhnhenn, U. Weinand and F. Wulf, Nucl. Instr. Methods in Phys. Research A 526 (2004) 537.

[7] R.J. Glauber and M. Lewenstein, Phys. Rev. A 43, (1991) 467.

[8] T. Okoshi, *Optical Fibers*, Academic Press, 1982

[9] L.S. Bogdankevich, B.M. Bolotovskii, J. Exp. Theoret. Phys. 32, 1421 [Sov. Phys. JETP 5, 1157] (1957).

[10] N.K. Zhevago, V.I. Glebov, Nucl. Instr. Methods A 331 (1993) 592; Zh. Exp. Teor. Fiz. 111 (1997) 466.

[11] A. Tomita and R. Chiao, Phys. Rev. Lett. 57 (1986) 937.

[12] J.D. Jackson, *Classical Electrodynamics*, John Wiley & Sons, Inc. (1962).

[13] W.K.H. Panofsky and M. Phillips, *Classical Electricity and Magnetism*, Addison-Wesley, Inc. (1962).

[14] T. Kalkbrenner, M. Ramstein, J. Mlynek and V. Sandoghdar, J. Microsc. 202 (2001) 72.

[15] P. Anger, P. Bharadwaj and L. Novotny, Phys. Rev. Lett. 96 (2006) 113002.

[16] N. K. Zhevago, Europhys. Lett. 15 (1991) 277.

[17] F.J. García de Abajo and A. Howie, Phys. Rev. Lett. 80 (1998) 5180.

[18] N. Yamamoto, K. Araya and F.J. García de Abajo, Phys. Rev. B 64 (2001) 205419.

[19] G. Naumenko, X. Artru, A. Potylitsyn, Y. Popov, L. Sukhikh and M. Shevelev, J. Phys.: Conf. Ser. 236 (2010) 012004.

[20] F.J. García de Abajo, Phys. Rev. Lett. 82 (1999) 2776; F.J. García de Abajo, Phys. Rev. E 61 (2000) 5743.

[21] D.W. Rule and R.B. Fiorito, AIP Conference Proceedings 229 (1991) 315.

[22] V.A. Lebedev, Nucl. Instr. Methods A 372 (1996) 344.

[23] J.-C. Denard, P. Piot, K. Capek, E. Feldl, Proc. of the 1997 Particle Accelerator Conference.
[24] X. Artru, R. Chehab, K. Honkavaara, A. Variola, Nucl. Instr. Methods B 145 (1998) 160.

7

Mechanical Properties of Optical Fibers

Paulo Antunes, Fátima Domingues, Marco Granada and Paulo André
Instituto de Telecomunicações and Departamento de Física, Universidade de Aveiro
Portugal

1. Introduction

Nowadays, optical communications are the most requested and preferred telecommunication technology, due to its large bandwidth and low propagation attenuation, when compared with the electric transmission lines. Besides these advantages, the use of optical fibers often represents for the telecom operators a low implementation and operation cost.

Moreover, the applications of optical fibers goes beyond the optical communications topic. The use of optical fiber in sensors applications is growing, driven by the large research done in this area in recent years and taking the advantages of the optical technology when compared with the electronic solutions. However, the implementation of optical networks and sensing systems in seashore areas requires a novel study on the reliability of the optical fiber in such harsh environment, where moisture, Na^+ and Cl^- ions are predominant.

In this work we characterize the mechanical properties, like the elastic constant, the Young modulus and the mean strain limit for commercial optical fibers. The fiber mean rupture limit in standard and Boron co-doped photosensitive optical fibers, usually used in fiber Bragg grating based sensors, is also quantify. Finally, we studied the effect of seawater in the zero stress aging of coated optical fibers. Such values are extremely relevant, providing useful experimental values to be used in the design and modeling of optical sensors, and on the aging performance and mechanical reliability studies for optical fiber cables.

2. Mechanical properties

The optical fibers are mainly used as the transmission medium in optical communications systems, nevertheless its applications in sensing technology is growing. Although the optical fiber mechanical properties are important for its use in optical communications (bending radius) is on the sensing applications that these properties are more relevant. In sensing technology the physical properties of the optical fiber are essential for the sensors characterization. Most of the optical fiber based sensors rely on the deformation of the optical fiber to determine the external parameter of interest. As an example, fiber Bragg grating (FBG) are one of the most promising technologies in sensing applications due to it numerous advantages, like small size, reduced weight, low attenuations, immunity to electromagnetic interference and electrical isolation (Antunes et al., 2011). The FBG concept as sensor relies on the mechanical deformation of the optical fiber to measure static or dynamic parameters like deformation, temperature or acceleration, therefore it is crucial to know the mechanical properties of the optical fiber (Antunes et al., 2008).

If an optical fiber is perturbed mechanically, it will suffer a deformation proportional to the amplitude of the perturbation force. This approach is valid for perturbations values lower than the elastic limit of the optical fiber, where the mechanical perturbations are reversible. The Hooke's law expresses the relation between the perturbation force and the produced deformation, the proportionality is given by the material elastic constant. The Hooke's law is given by the following expression, along the longitudinal axis of the fiber:

$$K = \frac{|F|}{|\Delta L|} \tag{1}$$

where K is the elastic constant and Δl is the relative deformation imposed by the action of the perturbation force, F.

The fiber Young modulus, E_G, is the proportionality constant between the perturbation force per area and the relative deformation:

$$F = E_G A \frac{\Delta l}{l} \tag{2}$$

In expression (2), A is area and l is length the optical fiber under perturbation. Considering expressions (1) and (2), the elastic constant is given by:

$$|K| = \frac{E_G A}{l} \tag{3}$$

According to expression (2), the slope of the linear region (elastic region) of the perturbation force as a function of the relative deformation represents the product $E_G \times A$. This product can be used in expression (3) to obtain the elastic constant of the optical fiber, knowing its length.

In this work we tested standard optical communications fiber SMF-28e from Corning, which according to the manufacturer specifications, have an uncoated diameter of 125.0±0.3 μm and 245±5 μm with the protective coating and photosensitive optical fiber PS1250/1500 co-doped fiber, from FiberCore, with 125±1 μm diameter without coating.

The strain measurements were made using a Shimadzu AGS-5kND mechanical test machine. The tested optical fibers were: ten samples of standard optical fiber with the acrilate protective coating; nine samples of standard optical fiber without the acrilate protective coating and ten samples of photosensitive optical fiber without the acrilate protective coating. Each sample had a length of 20.0 cm and was glued, with Cyanoacrylate glue, in each extremity to an aluminum plate to make it possible the fixation to the mechanical test machine.

Figure 1 shows photography of one optical fiber samples.

Fig. 1. Photography of a tested optical fiber with the glued aluminum plates on the extremities.

The experimental data collected with the mechanical test machine allows the representation of the force versus strain curves, for each sample of fiber.

Figure 2 shows the force versus strain graphic, representing the force applied along the fiber longitudinal axis, for each of the ten samples of standard optical fiber with the acrilate protective coating.

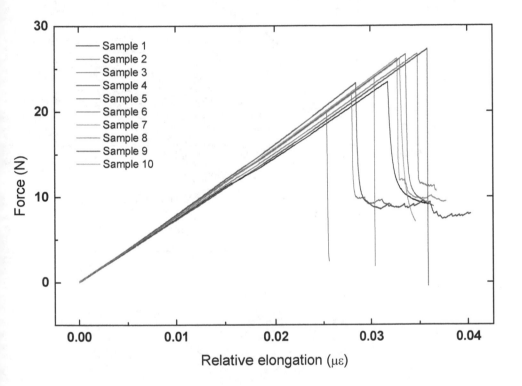

Fig. 2. Force versus strain experimental data for each sample of the standard optical fiber with the acrilate protective coating.

Considering the ten fiber samples of standard optical fiber with the acrilate protective coating, the average value for the product $E_G \times A$ is 780.87±24.99 N and the average rupture force value for the fibers fracture is 24.60±2.38 N.

The force versus strain curves for the standard fiber without the protective coating samples are presented in figure 3.

The average value obtained for the $E_G \times A$ product of the standard optical fiber without the coating was 849.42±6.88 N and the average rupture force was 4.35±1.45 N.

The force versus strain curve for the photosensitive optical fiber without the protective coating is presented in figure 4.

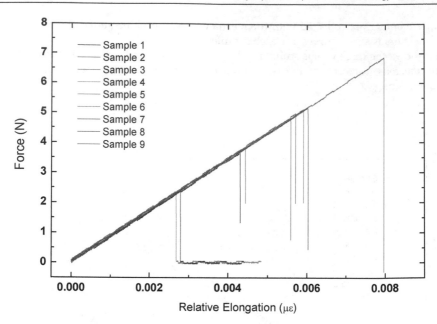

Fig. 3. Force versus strain experimental data for each sample of standard optical fiber without the acrilate protective coating.

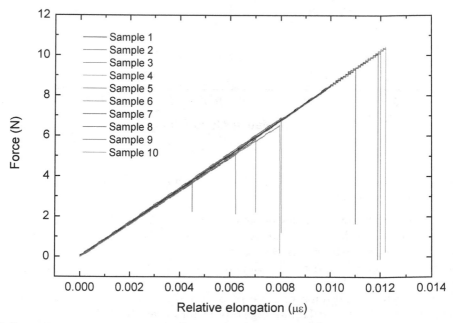

Fig. 4. Force versus strain experimental curve for the ten samples of photosensitive optical fiber without the acrilate protective coating.

For the photosensitive optical fiber without the coating the average value obtained for the $E_G \times A$ product was 841.30±15.06 N and the average rupture limit was 7.57±2.51 N.
Considering the area of each optical fiber type, the Young modulus can be obtained through the $E_G \times A$ product. The Young modulus for each type of optical fiber is presented in table 1, considering the average value of the $E_G \times A$ product.

Fiber	E_G (GPa)
Standard fiber with coating	16.56±0.39
Standard fiber without coating	69.22±0.42
Photosensitive fiber without coating	68.56±1.47

Table 1. Young modulus for different types of optical fiber.

The Young modulus values found in literature (Pigeon et al., 1992; Mita et al., 2000; Antunes et al., 2008) for silica and silica fibers are consistent with the ones we measure. The obtained values can be used in the design and modeling of optical fiber sensors were the fiber can be under some kind of stress, like FBG based sensors.

3. Optical fiber degradation behavior

The application of optical fiber in aggressive environments may lead to the degradation of its physic reliability and therefore the performance of the systems in which it is applied (El Abdi et al., 2010). Therefore, the conservation of the optical fiber physical characteristics in harsh environments, where the fiber is exposed to abrasion and moisture, is a key point to assure its good performance.
The optical fiber coatings can provide a robust protection from the extrinsic factors that may decrease its strength and performance. Nevertheless, and in spite of the protection provided by the coatings, the fiber is still permeable to moisture. There are reports of its ability to retain the hydroxyl groups from the water molecules,(Berger et al., 2003; Méndez et al., 2007; El Abdi et al., 2010; André et al., 2011), but also of the diffusion of other ions in addition to the ones from water. Such ions diffusion can be responsible for the optical fiber degradation and the decrease of its strength (Thirtha et al., 2002; Lindholm et al., 2004). Therefore, the behavior of the fiber strength with aging is not only dependent on the moisture present in the environment but also on the diffusion rate of the ions across the coating (Armstrong et al., 1999; Domingues et al., 2010).
The implementation of optical fiber systems (as for example in sensing structures or optical networks) (Ferreira et al., 2009), is dependent on the optical fiber reliability and its lifetime degradation in abrasive environments, where ions like Na+, Cl-, or even moisture are present.

3.1 Effect of maritime environment in the zero stress aging of optical fiber

To study the effect of maritime environment in the aging of optical fiber, several samples of a standard single mode fiber (SMF-28eR) fiber, manufactured by Corning, were left in a Sodium Chloride (NaCl) aqueous solution. The fiber under test had a diameter of 125 μm, d_f, and a total diameter of 250 μm with the coating, d_c. The effect of different NaCl concentrations were studied, namely, 0 (pure deionized water), 35, 100 and 250 g/L. The 35 g/L solution matches the average sea water concentration of NaCl, while the 250 g/L corresponds to the highest Sodium Chloride concentration found in the Earth, namely in the Dead Sea.

During the tests, samples were removed from the NaCl solution and its fracture stress was measured. For that procedure it was used an experimental setup like the one represented in figure 5.

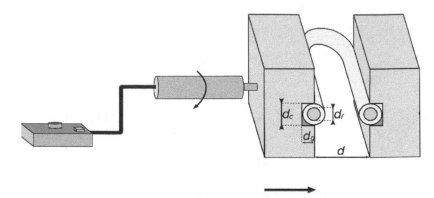

Fig. 5. Illustration of the experimental setup used to measure the fiber fracture stress.

The experimental setup uses a fixe PTFE plate and a moveable PTFE plate with grooves for the fiber fixing. The moveable plate is controlled by an electric translation stage (Newport, model 861). Initially, the plates have a distance between them of 20 mm. This distance is reduced at a speed of 0.55 mm/s. After the fiber break, the distance between plates is measured and related with the fiber fracture stress, which is dependent of the distance between the two plates. The stress applied in the fiber can be calculated through the equation (4)(El Abdi et al., 2010):

$$\sigma = E_0 \varepsilon \left(1 + \frac{1}{2} \alpha \varepsilon \right) \tag{4}$$

where σ is the stress in the fiber, E_0 the fiber young modulus, ε the strain in the fiber and α is a non linearity elastic parameter. The strain ε is given by (5):

$$\varepsilon = 1.198 \left[\frac{d_f}{d - d_c + 2d_g} \right] \tag{5}$$

where d_f is fiber diameter without the coating, d the distance between plates, d_c the fiber diameter with coating and d_g is the total depth of the two grooves. The non linearity elastic parameter, α, is given by:

$$\alpha = \frac{3}{4} \alpha' + \frac{1}{4} \tag{6}$$

For an optical fiber $\alpha'=6$ (El Abdi et al., 2010). By applying equation (4) to the measured data, we can determinate the applied stress at which the fiber fractures. This procedure was executed for the NaCl concentrations previously referred and for different aging periods.

For every sample the average stress values and error were calculated for five identical samples. In figure 6 is presented the results obtained along the aging time for the three concentrations under study.

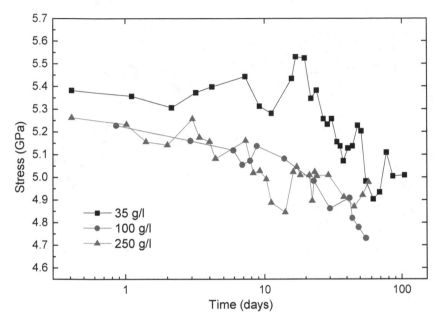

Fig. 6. Fracture stress values along degradation time for the fibers aged in the 35, 100 and 250 g/L NaCl solutions.

Several studies have reported that when the optical fiber is submitted to harsh environments its strength drastically drops after a certain time, showing a fatigue transition generally called as "knee". This behavior represents the sudden decrease of the optical fiber strength and the transition of its strength to a degradation regime (Armstrong et al., 1999; El Abdi et al., 2010). From the analyses of data collected is possible to observe that, for the fibers aged in the solutions with higher NaCl concentrations the strength of the fiber decreases and the "Knee" appears sooner in time.

In order to understand the stress behavior of aged fibers, the fracture stress values obtained for the three concentrations under study were fitted to a Boltzmann function, given by equation (7):

$$y = \frac{(A_1 - A_2)}{1 + \exp((x - x_0)/d_x)} + A_2 \qquad (7)$$

The parameters A1 and A2 are the upper and lower limit of the fiber stress, respectively, x_0 is the activation parameter. The d_x parameter represents the function higher slope and assumes, in that point, a value of $(A2-A1)/(4d_x)$.

In figure 7 is presented the behavior of the fracture stress of the fiber as function of the exposition time in a logarithmic scale to the 35, 100 and 250 g/L NaCl concentrated solutions and the fit to the Boltzmann function:

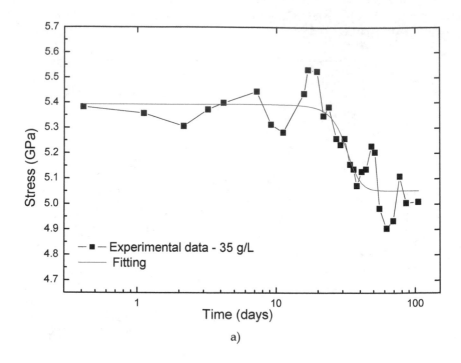

a)

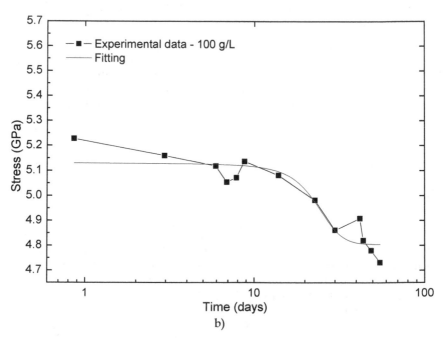

b)

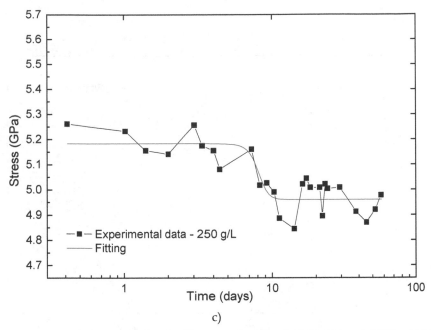

c)

Fig. 7. Fracture stress along time for the fibers aged in the a) 35, b) 100 and c) 250 g/L NaCl solutions and fitting to the Boltzmann function.

The table 2 displays the Boltzmann fit parameters for the three different concentrations.

Concentration (g/L)	A1 (GPa)	A2 (GPa)	x0 (days)	dx (days⁻¹)
35	5.39	5.05	31.34	3.97
100	5.13	4.80	23.13	4.30
250	5.18	4.96	8.05	0.65

Table 2. Boltzmann function fitting parameters.

From the values in table 2 we can see that, the time at which the strength transition occur (x_0) decreases with the increase of the NaCl concentration. If we establish a relation between the x0 value and the time at which the "knee" appears, we can affirm that the "Knee" will show up earlier for higher concentrations of NaCl.

This connection between the NaCl concentration and the time at which the strength transition occurs, is related to the ability of the ions in solution to infiltrate the fiber coating, react with the fiber glass surface, and remove the products through the coating (Thirtha et al., 2002). So, we can say that the diffusion through the coating of the species in solution or present in the environment in which the fiber is placed, has a major role on the aging behavior of the fiber, once such diffusion implies the decrease of the fiber strength and as consequence the decrease of its lifetime. In the analysis of the strength transition parameter, figure 8, we verify that the strength transition period has a degradation rate of 0.1174 days/[NaCl].

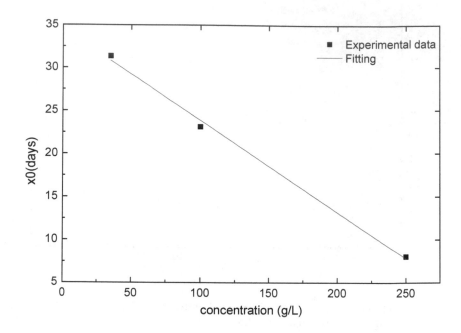

Fig. 8. Degradation rate of the aged optical fiber.

According to the authors (Danzer et al., 2007), the probability of fracture in a material increases with the number of flaws and with its dimension. Based on such assumption, the study of the probability to failure of the different aged samples will give us the information regarding the number of flaws in the samples.

To implement this study it was used the statistical Weibull's law given by:

$$\ln\left[\ln\left\{\frac{1}{1-F}\right\}\right] = m\left[\ln(\sigma) - \ln(\sigma_0)\right] \tag{8}$$

This law establishes the relation between the probability of fiber failure, F, with the applied stress σ. The first term represents the cumulative failure probability, and its evolution with the increase of the failure stress, ln(σ). The parameters σ_0 and m are constant, being m also referred to as the Weibull slope.

In the figure 9 it is represented the cumulative failure probability for fibers with no degradation and the ones aged in a 35 g/L solution of NaCl for 20 and 86 days.

It is possible to observe that the higher the aging period, the lower is the stress necessary to fracture the fiber.

Also for the different concentrations used and for the same aging periods, figure 10, it is clear that the higher the concentration of NaCl, the lower is the stress need for fracture to occur.

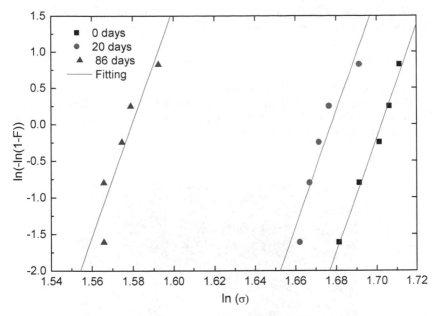

Fig. 9. Cumulative failure probability for fibers aged in the same NaCl concentration for different periods of time.

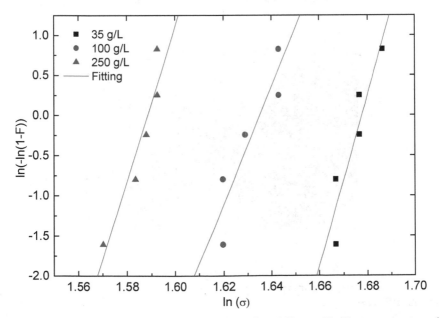

Fig. 10. Cumulative failure probability for fibers aged in different NaCl concentrations for the same period of time.

Through these latest analysis we may assume that the number of flaws in the fiber increases with the aging time and with the concentration of NaCl ions in solution.

3.2 Microscopic analysis of the aged optical fiber

In addition to the analytical study of the fibers, also its microscopic condition was analyzed through optical microscopy and SEM images.

In figure 11 we present the optical microscopy images, taken with an Olympus BH2 and the digital camera Sony DKC-CM30, for fibers degraded on a 35 g/l NaCl aqueous solution a) during 31 days and b) 105 days (Domingues et al., 2010).

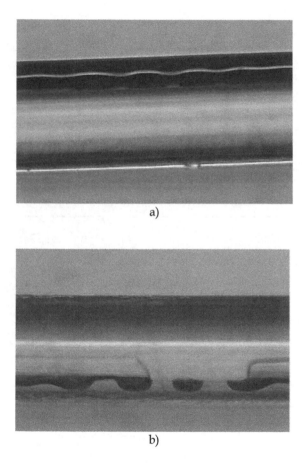

a)

b)

Fig. 11. Microscopy images from fibers aged in a 35 g/l NaCl aqueous solution a) during 31 days and b) 105 days.

In these images we can see the difference in the protective polymer as consequence of its degradation.

The SEM images were obtained with an Hitachi SU-70 apparatus after carbon evaporation. In figure 12, are represented some of the images collected, for the three concentrations under study.

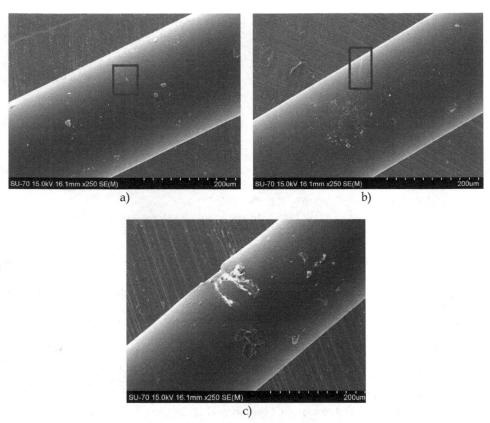

Fig. 12. SEM images from fibers aged in a) 35g/L NaCl solution for 105 days, b) 100g/L NaCl solution for 69 days and c) 250g/L NaCl solution for 64 days.

In these samples, it is possible to identify the damage induced in the coating in the three samples, being the most relevant, the one presented by the sample aged in a 250 g/L solution. But also in addition to the degradation, we can identify some NaCl crystal deposited in the fiber's surface.

4. Conclusions

We characterized the mechanical properties for commercial optical fibers. The Young modulus obtained has the value of 69.2 GPa for the standard optical fiber without the external acrilate protective coating. The effect of seawater in the zero stress aging of coated optical fiber, shows its increasing degradation with the Sodium Chloride concentration. The degradation rate has a value of 0.1174 days/[NaCl].

These results can be useful for the design and modeling of optical sensors, and on the aging performance of optical fiber deployed in telecommunication networks.

5. References

André, P. S., F. Domingues and M. Granada (2011). Impact of the Maritime Environment on the Aging of Optical Fibers, *Proceedings of CLEO2011: Laser Science to Photonic Applications*, Baltimore, USA.

Antunes, P., H. Lima, N. Alberto, L. Bilro, P. Pinto, A. Costa, H. Rodrigues, J. L. Pinto, R. Nogueira, H. Varum and P. S. André (2011). Optical sensors based on FBG for structural helth monitoring, in: *New Developments in Sensing Technology for Structural Health Monitoring*, S. C. Mukhopadhyay, Springer-Verlag.

Antunes, P., H. Lima, J. Monteiro and P. S. André (2008). Elastic constant measurement for standard and photosensitive single mode optical fibres, *Microwave and Optical Technology Letters*, Vol. 50, No. 9, pp. 2467-2469, ISSN: 1098-2760.

Armstrong, J. L., M. J. Matthewson, M. G. Juarez and C. Y. Chou (1999). The effect of diffusion rates in optical fiber polymer coatings on aging, *Optical Fiber Reliability and Testing*, Vol. 3848, pp. 62-69, ISSN: 0277-786X.

Berger, S. and M. Tomozawa (2003). Water diffusion into a silica glass optical fiber, *Journal of Non-Crystalline Solids*, Vol. 324, No. 3, pp. 256-263, ISSN: 0022-3093.

Danzer, R., P. Supancle, J. Pascual and T. Lube (2007). Fracture statistics of ceramics - Weibull statistics and deviations from Weibull statistics, *Engineering Fracture Mechanics*, Vol. 74, No. 18, pp. 2919-2932, ISSN: 0013-7944.

Domingues, F., P. André and M. Granada (2010). Optical Fibres Coating Aging induced by the Maritime Environment, *Proceedings of MOMAG2010*, Brasil.

El Abdi, R., A. D. Rujinski, M. Poulain and I. Severin (2010). Damage of Optical Fibers Under Wet Environments, *Experimental Mechanics*, Vol. 50, No. 8, pp. 1225-1234, ISSN: 0014-4851.

Ferreira, L. F., P. F. C. Antunes, F. Domingues, P. A. Silva, R. N. Nogueira, J. L. Pinto, P. S. Andre and J. Fortes (2009). Monitorization of Sea Sand Transport in Coastal Areas Using Optical Fiber Sensors, *2009 Ieee Sensors, Vols 1-3*, pp. 146-150

Lindholm, E. A., J. Li, A. Hokansson, B. Slyman and D. Burgess (2004). Aging behavior of optical fibers in aqueous environments, *Reliability of Optical Fiber Components, Devices, Systems, and Networks Ii*, Vol. 5465, pp. 25-32, ISSN: 0277-786X.

Méndez, A. and T. F. Morse (2007). *Specialty Optical Fibers Handbook*, Elsevier Inc.,

Mita, A. and I. Yokoi (2000). Fiber Bragg Grating Accelerometer for Structural Health Monitoring. Fifth International Conference on Motion and Vibration Control (MOVIC 2000). Sydney, Australia.

Pigeon, F., S. Pelissier, A. Mure-Ravaud, H. Gagnaire and C. Veillas (1992). Optical Fibre Young Modulus Measurement Using an Optical Method, *Electronics Letters*, Vol. 28, No. 11, pp. 1034-1035,

Thirtha, V. M., M. J. Matthewson, C. R. Kurkjian, K. C. Yoon, J. S. Yoon and C. Y. Moon (2002). Effect of secondary coating on the fatigue and aging of fused silica fibers, *Optical Fiber and Fiber Component Mechanical Reliability and Testing Ii*, Vol. 4639, pp. 75-81, ISSN: 0277-786X.

8

Non Linear Optic in Fiber Bragg Grating

Toto Saktioto[1,2] and Jalil Ali[1]
[1]Advanced Photonics and Science Institute, Faculty of Science,
Universiti Teknologi Malaysia, Skudai, Johor;
[2]University of Riau, Pekanbaru, Riau;
[1]Malaysia
[2]Indonesia

1. Introduction

A Fiber Bragg Grating (FBG) is a periodic variation of the refractive index of the core in the fiber optic along the length of the fiber. The principal property of FBGs is that they reflect light in a narrow bandwidth that is centered about the Bragg wavelength, λ_B (A. Orthonos and K. Kalli, 1999). FBGs are simple intrinsic devices that are made in the fibre core by imaging an interference pattern through the side of the fibre. They are used as flexible and low cost in-line components to manipulate any part of the optical transmission and reflection spectrum. FBG is formed by the periodic variations of the refractive index in the fiber core. Several techniques have been established to inscribe them with UV-lasers. However, these technologies are limited to photosensitive fiber core material, which are unsuitable for high power applications. Only recently modifications have been demonstrated in a non photosensitive fiber but at the expense of longer exposure times (K. W. Chow *et al.*, 2008). FBGs have all the advantages of an optical fibre, such as electrically passive operation, lightweight, high sensitivity with also unique features for self-referencing and multiplexing capabilities. This gives them a distinct edge over conventional devices (Nahar Singh *et. al*, 2006, Govind P. Agrawal 2002). Therefore, FBGs in optical fibers have a wide range of applications, such as for sensors, dispersion compensators, optical fibre filters, and all-optical switching and routing (T. Sun *et. al,*2002). An UV laser source is used to form FBG's in fiber optics either through internal writing or external writing technique (A. Orthonos *et al*, 1995). The novel idea of using soliton is introduced for FBG.

Solitons are particle-like waves that propagate in dispersive or absorptive media without changing their pulse shapes and can survive after collisions. Various types of optical soliton phenomenon have been studied extensively in the area of nonlinear optical physics. These include the nonlinear Schröedinger solitons in dispersive optical fibers, spatial and vortex solitons in photorefractive material, waveguides and cavity solitons in resonators (Y. S. Kivshar and G. P. Agrawal, 2003).

The principal objective of this topic is to investigate the soliton in FBG showing potential energy. The theory involved in the modelling of soliton is based on the coupled-mode theory including the Kerr nonlinearity, group velocity dispersion (GVD) and self phase modulation (SPM) The motion of a particle moving in FBG represents the pulse propagation in the grating structure of fiber optics exhibiting the existence of optical fiber. In order to describe the photon motion, the function of potential energy is depicted.

2. Properties of Fiber Bragg Grating

A simple form of Fiber Bragg Grating (FBG) in Figure 1 consists of a periodic modulation of the refractive index in the core of a single-mode optical fiber (Phing, H.S.*et al*, 2008) . These types of uniform fiber gratings, where the phase fronts are perpendicular to the fiber longitudinal axis with grating planes have a constant grating period, Λ.

$$\lambda_B = 2N_{eff}\Lambda \tag{1}$$

where Λ is the spatial period (or pitch) of the periodic variation and N_{eff} is the effective index for light propagating in a single mode fiber.

The Bragg condition is a manifestation of both energy and momentum conservation. Energy conservation requires that the frequency of the incident radiation and the reflected radiation is the same, means

$$\hbar\omega_f = \hbar\omega_i \tag{2}$$

Momentum conservation requires that the incident wave vector, k_i, plus the grating wave vector, K, equal the wave vector of the scattered radiation, k_f. This leads to an equation in which,

$$k_i + K = k_f \tag{3}$$

where the grating wave vector, K, has a direction normal to the grating planes with a magnitude $\dfrac{2\pi}{\Lambda}$. The diffracted wave vector is equal in magnitude, but opposite in direction to the incident wave vector.

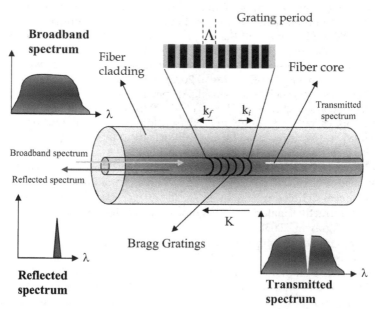

Fig. 1. A basic diagram of Fiber Bragg Grating (A. Orthonos and K. Kalli, 1999).

Hence, the momentum conservation becomes

$$2\left(\frac{2\pi n_{eff}}{\lambda_B}\right) = \frac{2\pi}{\Lambda}$$ (4)

Equation (4) simplifies to the first-order Bragg condition

$$\lambda_B = 2n_{eff}\Lambda$$ (5)

λ_B is the Bragg wavelength. This is the free space center wavelength of the input light that will be back-reflected from the Bragg grating region). n_{eff} is the effective refractive index of the fiber core at free space center wavelength.

3. Optical soliton in FBG

The existence of optical solitons in lossless fiber was theoretically demonstrated first by Hasegawa and Tappert in 1973. Bright and dark solitons appear in anomalous and normal dispersion regime respectively. The existence of an optical soliton in fibers is made by deriving the evolution equation for the complex light wave envelope via the slowly varying Fourier amplitude by retaining the lowest order of the group dispersion. This lower order is taken from the variation of the group velocity as a function of light frequency and the nonlinearity. For a glass fiber it is cubic and originates from the Kerr effect (K. Porsezian, 2007). The one soliton solution of the nonlinear Schrödinger equation is given by a sech T function which is characterized by four parameters, the amplitude, the pulsewidth, the frequency, time position and the phase. In particular, the soliton speed is a parameter independent of the amplitude unlike the case of Kortweg de Vries (KdV) soliton. This is important fact in the use of optical soliton as a digital signal. Originally in 1980, L. F. Mollenauer and his colleagues at Bell Laboratories succeeded in observing optical soliton in fiber. During the 1990's, many other kinds of optical soliton were discovered such as spatiotemporal solitons and quadratic solitons (Y. S. Kivshar and G. P. Agrawal, 2003).

Soliton in fibers is formed after the exact balancing of group velocity dispersion (GVD) arising as a combination of material and waveguide dispersion with that of the self-phase modulation (SPM) due to the Kerr nonlinearity. Due to this, a similar soliton-type pulse formation in Fiber Bragg Grating where the strong grating-induced dispersion is exactly counterbalanced by the Kerr nonlinearity through the SPM and cross-phase modulation (CPM) effects. As a result, there is a formation of slowly travelling localized envelope in FBG structures known as Bragg grating solitons. They are often referred to as gap solitons if their spectra lies well within the frequency of the photonic bandgap if the frequency of incident pulse matches the Bragg frequency. Thus based on the pulse spectrum with respect to the photonic bandgap, solitons in FBG can be classified into two categories as either Bragg grating solitons or gap solitons.

There are basically two conditions that one can determine the formation of solitons in FBG. First is based on high intensity pulse propagation in which the refractive index modulation is weak in FBG where nonlinear coupled-mode (NCM) equations are used to describe a coupling between forward and backward propagating modes. The other conditions deals with the low intensity pulse propagation in FBG where the peak intensity of the pulse is assumed to be small enough so that the nonlinear index change, n_2I is much smaller than the maximum value of δn. Under the low intensity limit, the NCM equations can be reduced to the nonlinear Schrödinger equation by using multiple scale analysis.

4. Coupled-mode theory for FBG

Several methods have been adopted to study and analyze the reflection and transmission properties of FBG (R. Kasyhap, 2004, M. Liu and P. Shum, 2006). The pulse propagation in FBG and its effect on Bragg grating affect the wave propagation in optical fibers can be examined using the coupled-mode theory (CMT) and Bloch wave technique. However, in this chapter we take CMT only into consideration.

One of the standard methods of analysis of FBG is using the coupled-mode theory (K. Thyagarajan and A. Ghatak and, 2007). According to this theory, the total field at any value of z can be written as a superposition of the two interacting modes and the coupling process results in a z-dependent amplitude of the two coupled modes. It is assumed that any point along the grating within the single-mode fiber has a forward propagating mode and a backward propagating mode. Thus the total field within the core of the fiber is given by

$$\Psi(x,y,z,t) = A(z)\psi(x,y)e^{i(\omega t - \beta z)} + B(z)\psi(x,y)e^{i(\omega t + \beta z)} \tag{6}$$

where x, y, z refers to space while t refers to variation of time, $A(z)$ and $B(z)$ represents the amplitudes of the forward and backward propagating modes (assumed to be the same order mode), $\psi(x,y)$ represents the transverse modal field distribution, ω refers to frequency and β is the propagation constant of the mode. The total field given by Equation (6) has to satisfy the wave equation given by

$$\nabla^2\Psi + k_0^2 n_g^2(x,y,z)\Psi = 0 \tag{7}$$

where $n_g^2(x,y,z)$ represents the refractive index variation along the fiber. For an FBG it is given by

$$n_g^2(x,y,z) = n^2(x,y) + \Delta n^2(x,y)\sin(Kz) \tag{8}$$

where $K = 2\pi / \Lambda$ represents the spatial frequency of the grating and Δn^2 represents the index modulation of the grating. For a uniform grating K is independent of z; when K depends on z, such gratings are referred to as chirped gratings. However, now we further focused on uniform gratings.

Substituting Equation (6) and Equation (7) into Equation (8) and making some simplifying approximations, we can obtain the following coupled-mode equations:

$$\frac{dA}{dz} = \kappa B e^{i\Gamma z} \quad \text{and} \quad \frac{dB}{dz} = \kappa A e^{-i\Gamma z} \tag{9}$$

where $\Gamma = 2\beta - K$ and κ represents the coupling coefficient given by

$$\kappa = \frac{\omega\varepsilon_0}{8}\iint \psi^* \Delta n^2(x,y)\psi \, dx dy \tag{10}$$

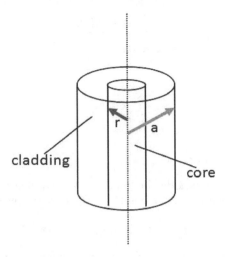

Fig. 2. Cross-section of an optical fiber with the corresponding refractive index profile (R. Kasyhap, 1999).

If the perturbation in the refractive index shown in Figure 2 is constant and finite only within the core of the fiber, then

$$\Delta n^2(x,y) = \Delta n^2, \quad r < a \tag{11}$$

$$= 0, \quad r > a$$

and we obtain

$$\kappa \approx \frac{\pi \Delta n}{\lambda_B} l \tag{12}$$

where λ_B is the Bragg wavelength and

$$l = \frac{\int_0^a \psi^2 r \, dr}{\int_0^\infty \psi^2 r \, dr} \tag{13}$$

The coupled-mode Equations (9) can be solved using the boundary conditions of

$$A\,(z = 0) = 1 \quad \text{and} \quad B\,(z = L) = 0 \tag{14}$$

where L is the length of the grating. Equation (14) implies that the incident wave has unit amplitude at $z = 0$ and the amplitude of the reflected wave at $z = L$ is zero because there is no reflected wave beyond $z = L$. We defined the reflectivity of the FBG by the ratio of the reflected power at $z = 0$ to the incident power at $z = 0$. Solving the coupled-mode equations and using the boundary conditions we obtain the reflectivity of the grating as follows:

$$R = \frac{\kappa^2 \sinh^2(\Omega L)}{\Omega^2 \cosh^2(\Omega L) + \frac{\Gamma^2}{4} \sinh^2(\Omega L)} \tag{15}$$

where

$$\Omega^2 = \kappa^2 - \frac{\Gamma^2}{4}$$

5. Pulse propagation in FBG

Wave propagation in a linear periodic medium has been studied extensively using coupled-mode theory. In the case of a dispersive nonlinear medium, the refractive index is given as

$$\bar{n}(\omega,z,I) = \bar{n}(\omega) + n_2 I + \delta n_g(z) \tag{16}$$

where n_2 is the Kerr coefficient and $\delta n_g(z)$ accounts for periodic index variation inside the grating. The coupled-mode theory can be generalized to include the nonlinear effects if the nonlinear index change, $n_2 I$ in Equation (2.11) is so small that it can be treated as a perturbation.

The starting point consists of solving Maxwell's equations with the refractive index given in Equation (16). When the nonlinear effects are relatively weak, we can work in the frequency domain and solve the Helmholtz equation,

$$\nabla^2 \tilde{E} + \tilde{n}^2(\omega,z)\omega^2 / c^2 \tilde{E} = 0 \tag{17}$$

The forward and backward propagating modes in FBG due to Bragg reflection can be described using CMT as been explained by Yariv in the distributed feedback structure (K. Senthilnathan, 2003). As usual, the governing equations for the pulse propagation in FBG are derived using Maxwell's equation. In this study the focus is on the frequency domain as the nonlinear effects are assumed to be relatively weak. It can easily be shown that Maxwell's equation are reduced to the following wave equation in the form

$$\frac{\partial^2 \vec{E}}{dz^2} - \frac{\varepsilon(z)}{c^2} \frac{\partial^2 E}{\partial t^2} = 0 \tag{18}$$

where perturbed permittivity, $\varepsilon(z) = \bar{n}^2 + \tilde{\varepsilon}(z)$, $\bar{n}^2$ is the spatial average of $\tilde{\varepsilon}(z)$, and $\bar{n}$ is the average refractive index of the medium. We consider the term $\tilde{\varepsilon}(z)$ with a period Λ and define $k_0 = \pi / \Lambda$. Using the Fourier series, $\tilde{\varepsilon}(z)$ can be written as

$$\tilde{\varepsilon}(z) = 2\tilde{\varepsilon}\cos(2k_0 z) \tag{19}$$

This electric field inside the grating can be written as

$$\vec{E}(z,t) = \vec{E}_f(z,t)e^{+i(k_b z - \omega_d t)} + \vec{E}_b(z,t)e^{-i(k_a z - \omega_a t)} + \dots \tag{20}$$

where $E_{f,b}(z,t)$ represents the forward and backward propagating waves, respectively, inside the FBG structure. Now, inserting Equation (19) and (20) into Equation (18) and considering that the fields $E_{f,b}(z,t)$ are varying slowly with respect to ω_0^{-1} in time and k_0^{-1} in space, the resulting frequency domain coupled mode equations can be written as

$$i\frac{\partial \vec{E}_f}{\partial z} + i\frac{\bar{n}}{c}\frac{\partial \vec{E}_f}{\partial t} + \kappa\vec{E}_b = 0 \tag{21}$$

$$-i\frac{\partial \vec{E}_b}{\partial z} + i\frac{n}{c}\frac{\partial \vec{E}_b}{\partial t} + \kappa\vec{E}_f = 0$$

The value of κ represents the coupling between the forward and backward propagating waves in the FBG. The set of Equations (21) are called linear coupled-mode (LCM) equations in which the non-phase-matched terms have been neglected. The LCM equations assume slowly varying amplitudes rather than the electric field itself. Note that CMT is an approximate description that is valid for shallow gratings and for wavelength close to the Bragg resonance.

6. Potential energy distribution in FBG

In the presence of Kerr nonlinearity, using CMT, the NLCM equations can be written as

$$i\frac{\partial \vec{E}_f}{\partial z} + i\frac{\bar{n}}{c}\frac{\partial \vec{E}_f}{\partial t} + \kappa\vec{E}_b + \left(\Gamma_s\left|\vec{E}_f^2\right| + 2\Gamma_x\left|\vec{E}_b^2\right|\right)\vec{E}_f = 0$$

$$-i\frac{\partial \vec{E}_b}{\partial z} + i\frac{\bar{n}}{c}\frac{\partial \vec{E}_b}{\partial t} + \kappa\vec{E}_f + \left(\Gamma_s\left|\vec{E}_b^2\right| + 2\Gamma_x\left|\vec{E}_f^2\right|\right)\vec{E}_b = 0 \tag{22}$$

where E_f and E_b are the slowly varying amplitudes of forward and backward propagating waves, $\bar{n}$ is the average refractive index, and Γ_s and Γ_x are SPM and Cross-Phase modulation terms. In Equation (22) the material and waveguide dispersive effects are not included due to the dispersion arising from the periodic structures dominates the rest near Bragg resonance condition. Noted that the above NLCM equations are valid only for wavelengths close to the Bragg wavelength.

Now, by substituting the stationary solution to the above coupled-mode equations is by assuming

$$E_{(f,b)}(z,t) = e_{(f,b)}(z)e^{-i\hat{\delta}ct/\bar{n}} \tag{23}$$

where $\hat{\delta}$ is the detuning parameter. Using the stationary solution in Equation (22), we obtain

$$i\frac{de_f}{dz} + \hat{\delta}e_f + \kappa e_b + \left(\Gamma_s\left|e_f\right|^2 + 2\Gamma_x\left|e_b\right|^2\right)e_f = 0$$

$$-i\frac{de_b}{dz} + \hat{\delta}e_b + \kappa e_f + \left(\Gamma_s\left|e_b\right|^2 + 2\Gamma_x\left|e_f\right|^2\right)e_b = 0 \tag{24}$$

Equation (24) represents the time-independent light transmission through the grating structure, where e_f and e_b are the forward and backward propagating modes κ represents n_{0k}, $\left(n_{0k} = \dfrac{n_{01} - n_{02}}{\pi} \right)$ where n_{01} is the core refractive index and n_{02} is the cladding refractive respectively, Γ_s represents Self Phase Modulation and Γ_x represents Cross-phase modulation effects. This has been extensively investigated by many researchers. The NLCM equations are non-integrable in general. But in a few cases, NLCM equations have exact analytical solutions representing the solitary wave solutions. However, Christoudolides and Joseph have obtained the soliton solution to the NLCM equation, known as slow Bragg soliton, under the integrable massive Thirring model where the SPM and detuning parameter is set to zero. After using suitable transformation, it is used in nonlinear optics as a simple model to explain the self-induced transparency effect. Using the Stokes parameters they derived the relation of energy density for the stationary solution for the NLCM equation in terms of the Jacobi elliptic function.

There are some possible interesting soliton-like solutions apart from these stationary solutions. In the fiber Bragg grating, these soliton-like solution for the NLCM equations carry a lot of practical importance.

7. Solution of optical soliton using NLCM

Wave propagation in optical fibers is analyzed by solving Maxwell's Equation with appropriate boundary conditions. In the presence of Kerr nonlinearity, using the coupled-mode theory, the nonlinear coupled mode equation is defined under the absence of material and waveguide dispersive effects. The dispersion arising from the periodic structure dominates near Bragg resonance conditions and it is valid only for wavelengths close to the Bragg wavelength.

In order to explain the formation of Bragg soliton, consider the Stokes parameter since it will provide useful information about the total energy and energy difference between the forward and backward propagating modes. In this study, the following Stokes parameter are considered where

$$
\begin{aligned}
A_0 &= \left| e_f \right|^2 + \left| e_b \right|^2, \\
A_1 &= e_f e_b^* + e_f^* e_b, \\
A_2 &= i \left(e_f e_b^* - e_f^* e_b \right) \\
A_3 &= \left| e_f \right|^2 - \left| e_b \right|^2
\end{aligned}
\quad \text{and} \tag{25}
$$

with the constraint A_0^2 equals to the sum of $A_1^2 + A_2^2 + A_3^2$. In the FBG theory, the nonlinear coupled-mode (NLCM) equation requires that the total power $P_0 = A_3 = \left| e_f \right|^2 - \left| e_b \right|^2$ inside the grating is constant along the grating structures. Rewriting the NLCM equations in terms of Stokes parameter gives

$$
\frac{dA_0}{dz} = -2\kappa A_2, \quad \frac{dA_1}{dz} = 2\hat{\delta} A_2 + 3\Gamma A_0 A_2'
$$

$$\frac{dA_2}{dz} = -2\hat{\delta}A_1 - 2\kappa A_0 - 3\Gamma A_0 A_1, \quad \frac{dA_3}{dz} = 0 \tag{26}$$

In Equation (26), we drop the distinction between the SPM and cross modulation effects. Hence Equation (26) becomes $3\Gamma = 2\Gamma_x + \Gamma_s$. It can be clearly shown that the total power, P_0 (=A_3) inside the grating and is found to be constant meaning it is conserved along the grating structure. In the derivation of the anharmonic oscillator type equation, it is necessary to use the conserved quantity. This is obtained in the form $\hat{\delta}A_0 + \frac{3}{4}\Gamma A_0^2 + \kappa A_1 = C$, where C is the constant of integration and $\hat{\delta}$ is the detuning parameter. Equation (27) can further be simplified to (Yupapin, P.P. et al, 2010),

$$\frac{d^2 A_0}{dz^2} - \alpha A_0 + \beta A_0^2 + \gamma A_0^3 = 4\hat{\delta}C \tag{27}$$

where $\alpha = 2\left[2\hat{\delta}^2 - 2\kappa^2 - 3\Gamma C\right]$, $\beta = 9\Gamma\hat{\delta}$ and $\gamma = \frac{9}{4}\Gamma^2$. Equation (27) contains all the physical parameter of the NLCM equation. Physically, a represents the function of detuning parameters, and phase modulation factors (SPM and CPM). β represents the function of phase modulation factors (SPM and CPM) and the detuning parameters. Lastly, γ represents the phase modulation factor (SPM and CPM). In general, a, β and γ are the oscillation factors.

8. Potential energy distribution in FBG structures

In order to describe the motion of a particle moving within a classical anharmonic potential, we have the solution of Equation 28 in the form of

$$V(A_0) = -\alpha \frac{A_0^2}{2} + \beta \frac{A_0^3}{3} + \gamma \frac{A_0^4}{4} \tag{28}$$

It represents the potential energy distribution in a FBG structures while the light is propagating through the grating structures.

In Equation (28), β is not considered due to power conservation along the propagating of this FBG structure. The qualitative aspects of the potential well will change if the nonlinearity parameter of the wave equation is varied.

Figure 3 depicts the double-well potential under Bragg resonance condition where $\beta = 0$, $\gamma = 0.23$ and α is varied from 0.1 to 1.0. Photon with power of less than the total power, P_0 will only travel inside the well unless their energy exceeds the energy level. This would allow the photon to move outside the well.

Figure 4 explains the optimized point for various values of α. The graph clearly shows that the optimized points decreased exponentially when values of α are increased. However, when $\alpha \gg 1$, the trend of the curve is no longer valid since it turns into an almost linear relationship.

Figure 5 shows the motion of photon in double well potential under different values of gamma for the Bragg resonance condition of γ from 0.13 to 0.53. Note that the increment of gamma which is between $0.53 < \gamma < 1$ will reduce the double well potential to a single well potential.

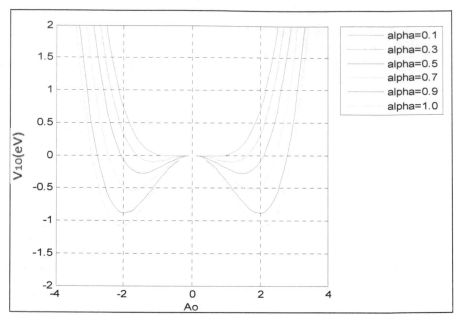

Fig. 3. The motion of photon in double well for different values of α.

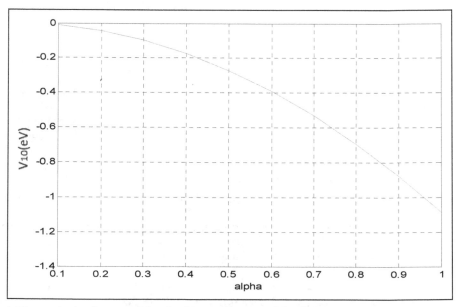

Fig. 4. The optimized point of the double well potential for different values of α.

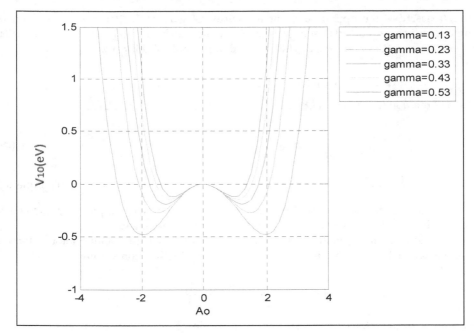

Fig. 5. Under Bragg resonance condition the system possesses double well potential for γ 0.13 to 0.53.

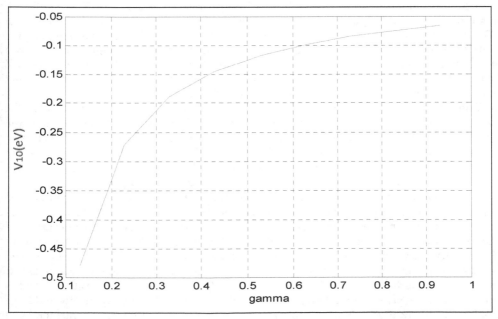

Fig. 6. The optimized point of the double well potential when γ = 0.1 to 1.0.

Figure 6 describes the optimized point for varies of gamma, γ. Parametric variation of gamma produces a potential energy function which increases exponentially. However, when $\gamma \gg 1$, a plateau is observed. This shows that it is not valid if $\gamma \to \infty$.

9. External disturbance of potential energy photon in FBG

By using Equation (28) where another term is considered, then we can have

$$\frac{d^2 A_0}{dz^2} - \alpha A_0 + \beta A_0^2 + \gamma A_0^3 + \theta A_0^4 = 4\hat{\delta}C \tag{29}$$

where $\alpha = 2\left[2\hat{\delta}^2 - 2\kappa^2 - 3\Gamma C\right]$, $\beta = 9\Gamma\hat{\delta}$, $\gamma = \frac{9}{4}\Gamma^2$ and $\theta = f(\theta)$. To simplify Equation (29), it is assumed the parameters of α, β and γ is independent with respect to parameter θ. Equation (29) contains all the physical parameter of the NLCM equation.

In order to describe the motion of a particle moving with the classic anharmonic potential, where the external disturbance is involved then we have the solution as follows,

$$V(A_0) = -\alpha \frac{A_0^2}{2} + \beta \frac{A_0^3}{3} + \gamma \frac{A_0^4}{4} + \theta \frac{A_0^5}{5} \tag{30}$$

It represents the potential energy distribution in the Fiber Bragg Grating structures.

Figure 7 depicts the motion of photon in a potential well which changes when few nonlinear parameters are taken into account as shown is Equation (30). Photon is trapped by the α parameter which is depicted by legend V. When α is too large, the potential well produces A_0 increases and have a wider of double well. The γ parameter is shown by X legend. When γ is large, the potential well produces A_0 increases. Suppose that the source is imposed to FBG than initial power is used to generate the particles. It shows that double well potential well is not symmetrical and the potential energy will decrease at certain region and is shown in Figure 7 in legend Y. The other effect is the disturbance at potential energy by legend Z where photon cannot be trapped symmetrically. It will tend to equilibrium but it is not stable where the photon leaves the potential curve as a losses.

In terms of parametric function, we can describe it as follows. The change in α will affect the dip of the potential well. If α is approximately too small, the shape of the potential well develop into a single potential well. The occurrence of β effect in the motion of photon gives an effect to the negative region which means $A_0 < 0$. The effect of γ also shows that the width of potential well will decrease if we increase the value of γ. Therefore if we increase the value of gamma, we can assume that the photon will be localized and can be trapped. In addition, another nonlinear factor θ, it will change the shape of potential well rapidly. We could say that if we include the existence of θ, the shape of potential well becomes chaotic. The photon does not only move within a certain region that is known as the potential well and moving freely.

Figure 8 shows the effect of external disturbance, θ. It shows that by increasing the value of θ, it will also affect the change in γ. In other words, the negative part of A_0 will be influenced it potential energy. The different values of γ will produce different profiles. By simulating, we assumed that the increased of γ value from 0.3 to 0.9, the curve will be positioned within the region C. The peak of V for each γ from 0.3 to 0.9 describes θ increases linearly and large gradient compare to the initial V. This represents that potential energy cannot maintain photon to be trapped and equilibrium state if γ is relatively small.

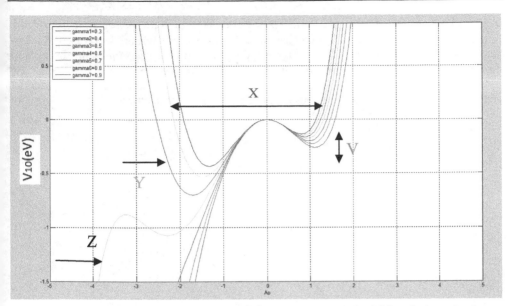

Fig. 7. The motion of photon in potential well for $a = 0.9$, $\beta = 0.3$, $\theta = 0.09$ and γ is varies from 0.3 to 0.9.

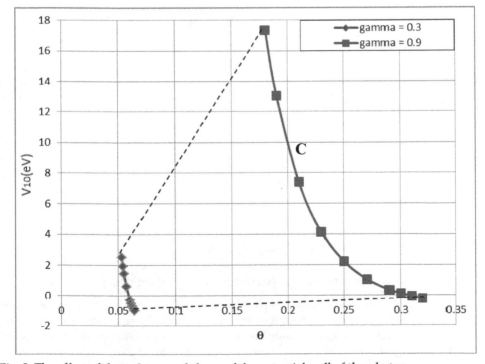

Fig. 8. The effect of theta, θ to γ and shape of the potential well of the photon.

In Figure 9 it can be shown that by increasing the value of β the potential energy of the potential well will be reduced. The highest potential drop occurs within the range of β, 0.2 to 0.3. If the disturbance is large, it requires a high potential energy to maintain the photon especially for $\gamma = 0.7$. In other words, increasing the γ value will affect the shape of the potential well in terms of the potential energy. It will affect the equilibrium of the potential well and therefore the trapped photons are no longer being trapped or localized.

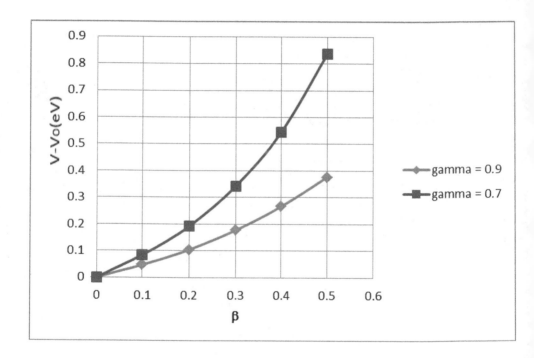

Fig. 9. The disturbance to the potential energy by β factor.

10. Photon due to external energy perturbation in potential well

Figure 10 depicts the motion of photon in potential well which changes when few nonlinear parameters is take into account as described by Equation (30). There are theoretically some comments in this figure. Photon is trapped by α parameter which is depicted by legend V. When α is too large, the potential well produces A_0 increases and have a wider double well. The value of γ parameter is shown by X legend. When γ is large, the potential well produces

A_0 increases. Suppose that the source is imposed to FBG than initial power is used to generate the particles. It shows that double well potential well is not symmetric and potential energy will decrease at the certain region in legend Y. The other effect is the perturbation of potential energy by legend Z where photon cannot be trapped symmetrically. It will tend to equilibrium but it is not stable where it can go for losses.

The change in the parametric function can be easily described in terms of a, β and γ. The dip in the potential well will transform with a single potential well when a is extremely small. β affects the photon motion which in turn will effect to the negative region of the potential well when $A_0 < 0$. The effect of γ shows that the width of potential well will decrease if we increased the value of γ. The photon will be trapped when γ is increased. The shape of the potential well can be controlled by a nonlinear factor θ. The changes in the value of θ lead to a chaotic behaviour of the potential well. Under these conditions the photon can either move within certain specific regions or act as a free particle. Thus, Figure 10 illustrates the single perturbation as described by the nonlinear parameter, θ.

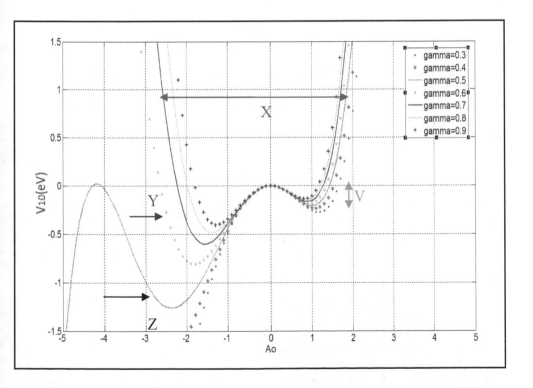

Fig. 10. The motion of photon in potential well for $a = 0.9$, $\beta = 0.3$, $\theta = 0.09$ and γ is varies from 0.3 to 0.9.

Now consider the case in Equation (27) with a set of constraints which is governed by $\phi_{(e)} = \sum\limits_{n=0}^{\infty} A_0{}^n$. The perturbation factor then is

$$\frac{d^2 A_0}{dz^2} = \phi_e{}''\big|_{n=0} \tag{31}$$

If Equation (30) is accumulated using the external perturbation then

$$\phi''\big|_{n=0} + \sum\limits_{\substack{n=0 \\ m=1}}^{\infty} C_m A_0{}^n = \psi$$

where ψ is a function of $f(\delta, C, C_m)$ and $C_m = [\alpha, \beta, \gamma, ...]$
The value of $m = 2n$ for $n = 1, 2, 3, ..., m = 2n + 1$ for $n = 0, 1, 2, ...$
C is constant and $C = (C_1, C_2, C_3, ..., C_m)$. The value of C is linear to A_0 but not to V. Equation (31) can then be modified by

$$V(A_0) = \sum\limits_{\substack{m=1 \\ n=0}}^{\infty} C_m A_0^n \tag{32}$$

Equation (32) represents the complete potential energy distribution in the Fiber Bragg Grating structure. We believe at this juncture, the potential function is modified from Conti and Mills (C. Conti and S. Trillo, 2001). Using well-known Duffing oscillator type equation, analogically it is written as

$$\phi_e{}'' + \sum\limits_{\substack{m=1 \\ n=0}}^{\infty} C_m A_0{}^n = 0 \tag{33}$$

For multi perturbation of nonlinear parameters, two major shapes will be simplified in the series term. The coupled mode equations are solved under different conditions when soliton is used for FBG writing. The cases examined are (i) when there is no energy disturbance (ii) the effect of potential energy disturbance factor (iii) potential energy with the highest disturbance factor.

When multi perturbations are considered then the photon will be trapped and untrapped for various conditions. As depicted in Figure 11, it explains the extrapolation of the graph if more factors of perturbation added into Equation (32). The addition of parametric factors by the higher odd number, Figure 11 (a) will allow the photon to move in a well, and Figure 11 (b) will lead the photon to be untrapped and higher even number. It is clearly shown in the graphs that as $n >> \infty$, the value of $|A_0|$ will remain constant in the range of $-2 < A_0 < 2$. However, when the value of $V_{(0)}$ is equal to zero, there are many possibilities of A_0, meaning the exact value of intentsity, A_0 to trap the photon is difficult to determine in this condition. If the parametric factor considered is too large then we may conclude that the photon is in indifferent state part of the equlibrium.

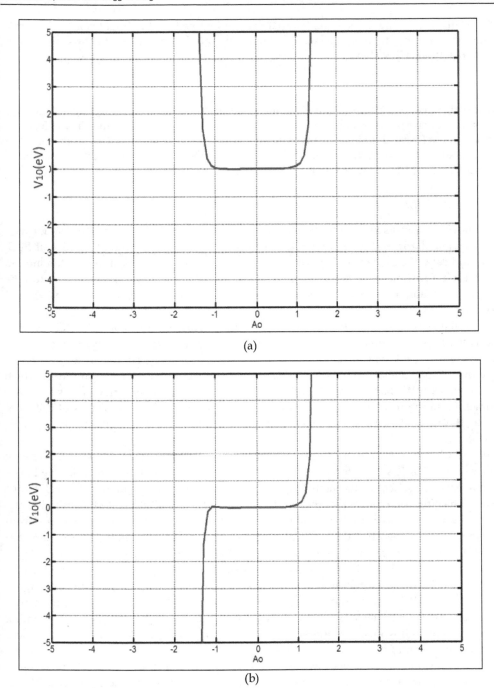

Fig. 11. The disturbance factor that affect the shape of the potential well of the motion of photon.

The stationary solutions of Equation (32) are applied neither for bright nor dark soliton solution since the dominant parametric factor in contributing A_0 is unknown. However, from Equation (32) we have

$$A_0 = f(C_m, z) \tag{34}$$

Under these conditions, the frequencies with photonic band gap keep forming an envelope after the exact balancing at grating-induced dispersion with nonlinearity. Either decay or increase, the forward and backward waves are transferred by Bragg reflection process. The total energy of the system, potential energy function is equal to zero having multi perturbation which is $-1 < A_0 < 1$ and if $V \to \infty$, $A_0 = 2$.

11. Conclusion

The novel idea of using soliton in Fiber Bragg Gratings (FBG) shows that the motion of a particle moving in FBG represents the pulse propagation in the grating structure of FBG. This indicates the existence of optical soliton. It is described in terms of the photon motion and as a function of potential energy. Results obtained show that photon can be trapped by nonlinear parameters of potential energy which are identified as α, β, γ and θ.

In the first simulation results of nonlinear parametric studies of photon in a FBG, we have successfully shown that the changes of nonlinearity parameter will affect the motion in the potential well. This will influence the existence of Bragg soliton in a fiber Bragg grating. In the second simulation results, we have added new nonlinear parameter which is known as θ. We have preset the value of θ, α and β and vary the value of γ over certain range. From the results, it is depicted that the factor θ will affect the shapes of potential well. If the existence of θ is taken into account, the potential well profile becomes chaotic.

The simulation data are then expanded on the multi perturbation of potential energy photon in FBG. It shows that the change of α affect the dip of the potential well. The occurrence of β effect in the motion will affect the soliton propagation in the region for $A_0 < 0$. The effect of γ shows that the width of potential well will decreased if the value of γ is increased. However, another nonlinear factor, θ will turns the shape of potential well rapidly which necessities the multi perturbation studies. When multi perturbations are considered, the photon will be trapped and entrapped under various conditions. From this, we may conclude the addition of nonlinear parametric factors by the higher odd number will allow the photon to move in a well instead to be entrapped with the higher even number. It is found from this study that the potential well under Bragg resonance condition is not symmetrical and conserved. The higher perturbation series representing the potential well is much indifferent of the equilibrium in both odd and even nonlinear parametric factor of n.

As a conclusion, these studies have successfully shown that it is plausible to use soliton for FBG writing and the soliton can be controlled by manipulating the parametric effects which are α, β, γ and θ. The model developed in this topic can be further extended by optimizing the nonlinear parameters in terms of the potential energy, soliton trapping and its applications as optical tweezers. The model can be tested by developing compact miniature FBG inscribing system using laser diode.

12. Acknowledgment

We would like to thank the Institute of Advanced Photonic Science, Faculty of Science, Universiti Teknologi Malaysia (UTM) and Physics Department, Faculty of Math and Natural Sciences, University of Riau, Pekanbaru, Indoneseia for generous support in this research.

13. References

Andreas Orthonos and Kyriacos Kalli. (1999)."Fiber Bragg Gratings – Telecommunications and Sensing". Artech House Boston, London.

Andreas Orthonos and Xavier Lee. "Novel and Improved Methods of Writing Bragg Gratings with Phase Masks". IEEE Photonics Technology Letters, Vol. 7, No. 10, October 1995.

C. Conti and S. Trillo, "Bifurcation of gap solitons through catastrophe theory". Phys. Rev. E 64 (2001), 036617.

Govind P. Agrawal. (2002). "Fiber Optic Communication System". Wiley-Interscience, U.S.A.

K. Senthilnathan et al. "Grating solitons near the photonic bandgap of a fiber Bragg grating," ScienceDirect, Chaos, Solitons and Fractals 33 (2007) 523-531.

K. Thyagarajan and Ajoy Ghatak. (2007). "Fiber Optic Essentials". Wiley-Interscience, Canada.

K. W. Chow, Ilya M. Merhasin, Boris A. Malomed, K. Nakkeeran, K. Senthilnathan and P. K. A. Wai. "Periodic waves in fiber Bragg gratings". Physical Review E, 77, 026602 (2008).

M. Liu and P. Shum. "Simulation of soliton propagation in a directional coupler," Springer, Optical and Quantum Electronics (2006) 38:1159-1165.

Nahar Singh, Subhash C. Jain, Vandana Mishra, G. C. Poddar, Palvinder Kaur, Himani Singla, A. K. Aggrawal and Pawan Kapur. "Fibre Bragg grating – based sensing device for petrol leak detection". Current Science. Vol. 9, No. 2, January 2006.

Phing, H.S., Ali, J., Rahman, R.A., Saktioto. "Growth dynamics and characteristics of fabricated Fiber Bragg Grating using phase mask method". Microelectronics Journal. (2008).

R. Kashyap, Fiber Bragg Gratings (Academic Press, San Diego, 1999)

T. Sun, S. Pal, J. Mandal, K. T. V. Grattan. "Fibre Bragg grating fabrication uysing fluoride excimer laser for sensing and communication applications". Central Laser Facility Annual Report 2001/2002.

Yuri S. Kivshar and Govind P. Agrawal. (2003). "Optical Solitons: From Fiber to Photonics Crystal". Academic Press, U.S.A.

Yupapin, P.P., Saktioto, T., Ali, J., Photon trapping model within a fiber bragg grating for dynamic optical tweezers use, *Microwave and Optical Technology Letters* 52 (4), pp. 959-961.(2010)

9

Optical Fibers and Optical Fiber Sensors Used in Radiation Monitoring

Dan Sporea[1], Adelina Sporea[1],
Sinead O'Keeffe[2], Denis McCarthy[2] and Elfed Lewis[2]

[1]*National Institute for Laser, Plasma and Radiation Physics, Laser Metrology Laboratory,*
[2]*Optical Fibre Sensors Research Centre, University of Limerick*
[1]*Romania,*
[2]*Ireland*

1. Introduction

By their very nature, optical fibers and, by extension, intrinsic and extrinsic optical fiber-based sensors are promising devices to be used in very different and complex environments considering their characteristics such as: capabilities to work under strong electromagnetic fields; possibility to carry multiplexed signals (time, wavelength multiplexing); small size and low mass; ability to handle multi-parameter measurements in distributed configuration; possibility to monitor sites far away from the controller; their availability to be incorporated into the monitored structure; wide bandwidth for communication applications. In the case of the optical fibers, the possibility to be incorporated into various types of sensors and actuators, free of additional hazards (i.e. fire, explosion), made them promising candidates to operate in special or adverse conditions as those required by space or terrestrial applications (spacecraft on board instrumentation, nuclear facilities, future fusion installations, medical treatment and diagnostics premises, medical equipment sterilization). Major advantages to be considered in using optical fibers/optical fiber sensors for radiation detection and monitoring refer to: real-time interrogation capabilities, possibility to design spatially resolved solutions (the capability to build array detectors), in-vivo investigations (i.e. inside the body measurements).

As information on the behavior and operation of optical fibers/optical fiber sensors under irradiation conditions are scattered over a great variety of journal papers and conference contributions dealing with many different fields (nuclear science and engineering; measurement science; material science; radioprotection; nuclear medicine and radiology; sensor design; radiation dosimetry; fusion installations concepts; particle accelerators; astrophysics and space science; defense and security; lasers, optics, optical fibers and optoelectronics; physics and applied physics; scientific instrumentation; radiation effects) we decided to design this book chapter as a comprehensive review on the subject. The chapter opens with some general considerations on the radiation–matter interaction, and continues with a review of irradiation effects on different types of optical fibers (silica optical fibers, plastic optical fibers, special optical fibers), effects which can be considered when radiation sensors are developed. The next issue addressed refers to environments where optical fibers/optical fiber sensors are employed for radiation monitoring/

dosimetry. The main part of the chapter is dedicated to the presentation of major proposed designs for intrinsic and extrinsic optical fiber sensors for radiation measurements. The experimental set-ups and irradiation conditions we used for the assessment of irradiation effects on various optical fibers are introduced and samples from our results illustrate the possible use of such optical fibers in radiation monitoring/dosimetry. The Romanian team focused its work on tests of silica and sapphire optical fibers subjected to different irradiation conditions (alpha particles, beta, gamma and X rays, neutron), while the Limerick group investigated radiation detectors based on plastic optical fibers, under gamma and X-rays irradiation.

Radiation induced effects in materials and devices are evaluated based on the energy losses resulting upon the interaction between highly energetic radiation and matter (Wrobel, 2005). Losses associated to these interactions can imply atoms ionization or can induce non-ionizing effects, such as changes of the vibrational/rotational states of molecules, atoms vibrations or atoms displacement. Generically, the term highly energetic radiation covers: a) uncharged particles – photons (gamma-rays, X-rays); b) charged light (electrons, positrons) and heavy (different types of ions) particles; c) nucleus constituents (neutrons, protons). Accelerated alpha particles (two protons and two neutrons bounded) or electrons, as well as high energy neutrons or photons can produce an atom's ionization. Depending on their energy, charge particles and nucleons can lead to either elastic or non-elastic interactions. On the other hand, lower energy photons can contribute to phenomena such as Rayleigh scattering (elastic scattering of photons by atoms or molecules), Compton scattering (the decrease of a X-ray or gamma-ray energy as it undergoes a non-elastic interaction with matter, the lost energy being transferred to a scattering electron as part of a ionizing process), photo-electric effect (the emission of electrons by a solid, liquid or gas upon the absorption of photons in that material) or pair production (the generation of an electron-positron pair as photons, having sufficient energy, interact with a nucleus). The occurrence of a particular phenomenon depends on the energy spectrum of the incident photons and the interaction cross sections (depending on the material involved). Light charged particles (i.e. electrons) when travelling inside a material are deflected by the surrounding atoms and lose energy, which is converted to photon energy, so, a continuum spectrum electromagnetic radiation is generated (Bremsstrahlung or deceleration radiation). Cerenkov radiation (a continuum spectrum optical radiation having a higher intensity at the UV wavelength end of the optical spectral range) is produced when a charged particle (i.e. electron) propagates within a dielectric medium with a speed higher than the phase velocity of the light in that medium. The molecules of the medium are polarized by the travelling particles and, returning to the ground state, emit visible optical radiation.

2. Optical fibers performances under irradiation

Exposed to ionizing radiation, silica optical fibers exhibit effects such as: radiation induced absorption (RIA), radiation induced luminescence (RIL), increase of the optical radiation scattering as it propagates over the fiber length, thermoluminescence, change of the waveguide refractive index. RIA and RIL effects contribute, generally speaking, to the degradation of the signal-to-noise ratio of the signal transmitted over the optical fiber guide. Attempts were made to reformulate the problem and to use these effects as a measure of the dose rate/total dose of the radiation to which the optical fiber is exposed.

In assessing the behavior of silica optical fibers under irradiation one has to consider several premises which have a major influence on the irradiation induced absorption, and hence on a would-be application of such optical fibers to radiation dosimetry. The RIA is affected by:

1. the manufacturing conditions, the parameters related to the technology used in producing the optical fiber: the deposition conditions, the draw process characteristics – draw speed, fiber drawing tension, the preform deposition temperature, oxygen-to-reagent ratio (02/R) used during core and clad deposition (Friebele, 1991; Girard et al., 2006; Hanafusa et al., 1986);

2. the existence, prior to the irradiation, of some precursors (Miniscalco et al., 1986);

3. the dopants present in the optical fiber core or cladding (pure silica, or doped with Ce, Er, Ge, F, N, P, Yb, high-OH, low-OH, high-Cl, low-Cl, H2-loading), (Arvidsson et al., 2009; Berghmans, 2006; Berghmans et al., 2008; Bisutti et al., 2007; Brichard & Fernandez Fernandez, 2005; Friebele, 1991; Girard et al., 2004a; Girard et al., 2004b; Girard et al., 2008; Griscom et al., 1996; Henschel et al., 1992; Kuyt et al., 2006; Lu et al., 1999; Mady et al., 2010; Paul et al., 2009; Regnier et al., 2007; Vedda et al., 2004; Wijnands et al., 2007), in some situations such ingredients contribute to the radiation hardening (Brichard et al., 2004; Brichard & Fernandez Fernandez, 2005; Girard et al., 2009);

4. the residual substances remaining after the manufacturing process, as for example chlorine having an associated color center in the UV spectral range, which can extend into the visible (Girard et al., 2008; Girard & Marcandella, 2010);

5. the type of radiation to which the optical fiber is subjected (Arvidsson et al., 2009; Bisutti et al., 2007; Brichard et al., 2001; Calderón et al., 2006; Girard et al., 2004a; Girard et al., 2004b; Girard et al., 2006; Girard et al., 2008; Merlo & Cankoçak, 2006; Toh et al., 2004; Yaakob et al., 2011);

6. the irradiation conditions: total dose, dose-rate, steady-state, pulsed, cyclic (Arvidsson et al., 2009; Berghmans, 2006; Berghmans et al., 2008; Kuhnhenn, 2005; Lu et al., 1999; Paul et al., 2009 ; Thériault, 2006; Wijnands et al., 2007);

7. the annealing effect and the temperature stress applied to the optical fiber during or post irradiation (Okamoto et al., 2004; Thériault, 2006; Toh et al., 2004; Tsuchiya et al., 2011);

8. the optical spectral band considered for the RIA evaluation (Regnier et al., 2007);

9. the optical fiber coating (Brichard, 2002; Gusarov et al., 2008b);

10. the photobleaching process and the level of the optical power injected into the optical fiber for measurement and its wavelength (Arvidsson et al., 2009; Miniscalco et al., 1986; Thériault, 2006; Treadaway et al.,1975).

The complexity of changes induced by highly energetic radiation in the optical transmission of silica optical fibers is governed by the optical fiber design, i.e. core-cladding dopants and diameter ratio (Deparis et al., 1997; Friebele, 1991; Girard et al., 2004a; Girard et al., 2004b; Kuhnhenn, 2005; Paul et al., 2009), and by the multitude of color centers activated under the irradiation process, such as (Brichard & Fernandez Fernandez, 2005; Origlio, 2009):

1. Ge centers: GeE', Ge(1), Ge(2), GEC, GeX, Ge-NBOH (Non-Bridging Oxygen Hole) (Alessi et al., 2010 ; Bisutti et al., 2007; Girard et al., 2006 ; Girard et al., 2008; Lu et al., 1999; Wijnands et al., 2007) ;

2. P centers: P1, P2, P4, Phosphorus-Oxygen-Hole - POHC (Bisutti et al., 2007; Girard et al., 2006; Girard et al., 2011 ; Paul et al., 2009; Wijnands et al., 2007) ;

3. silica related paramagnetic centers (Non-Bridging Oxygen Hole - NBOHC, PerOxy-Radical - POR, SiE', Self-Trapped Hole - STH 1/2) or diamagnetic centres (Oxygen

Deficient Centre – ODC; Peroxy Linkage – POL), (Berghmans et al., 2008; Girard et al., 2008).

In order to distinguish the contribution of various color centers to the irradiation induced absorption in silica optical fibers, complementary investigations, apart from the off-line/on-line optical transmission measurements, were carried-out: Electron Paramagnetic Resonance – EPR (Girard et al., 2008; Radiation effects, 2007; Sporea et al., 2010a; Weeks & Sonder, 1963), luminescence (Girard et al., 2005; Girard et al., 2006; Sporea et al., 2010a; Miniscalco et al., 1986), thermoluminescence (Espinosa et al., 2006; Hashim et al., 2008; Mady et al., 2010; Sporea et al., 2010b; Yaakob et al., 2011), confocal microscopy luminescence and Raman analysis (Girard et al., 2008; Origlio, 2009), time resolved photoluminescence (Cannas et al., 2008). Based on the results derived from different sources, a decomposition of the optical absorption spectra by using Gaussian bands associated to known color centers can be performed (Girard et al., 2008; Girard et al., 2011; Sporea et al., 2010a).

Apart from the aspects previously described (increase of the optical absorption in irradiated optical fibers, post irradiation luminescence and thermoluminescence effects) other phenomena have to be considered as prospective candidates for the use of optical fibers in radiation monitoring:

1. generation of Cerenkov radiation when optical fibers are subjected to a charged particle flux;
2. radiation induced luminescence (RIL) in optical fibers under irradiation (Skuja et al., 1996).

Traveling in a dielectric, transparent media, a charged particle (i.e. electron), having a velocity greater than the optical radiation phase velocity in that medium, interact with medium molecules by polarizing them. As the molecules return very fast to their ground state an optical radiation is emitted (Cerenkov, 1958). The optical radiation has a continuum distribution over the UV-visible spectral range.

As the emitted radiation spectrum varies with $1/\lambda^3$, the spectrum of the Cerenkov radiation is predominantly in UV-blue. Investigations on the generation of Cerenkov radiation were carried out for both silica and plastic optical fibers, such as polymethylmethacrylate - PMMA (Intermite et al., 2009; Jang et al., 2010a). The efficiency of coupling to and detecting the Cerenkov radiation with an optical fiber is a function of the type of optical fiber (its refractive index), the particles energy, the irradiation geometry: angle of incidence of the particle beam on the optical fiber axis, the optical fiber NA, the distance between the particle trajectory and the optical fiber axis (Goettmann et al., 2005; Jang et al., 2010a; Jang et al., 2011a).

In the case of a PMMA fiber with $n = 1.49$, for an electron beam energy of 6 MeV the maximum value for θ is 47.7⁰ (Jang et al., 2010a), while for a silica optical fiber ($n = 1.46$) and a energy of the electron beam of 175 keV the Cerenkov radiation angle was evaluated to be about 46⁰ (Intermite et al., 2009).

Gamma-rays can lead to the generation of Cerenkov radiation only if a Compton converter target is placed between the gamma source and the optical fiber. Photons incident on the converter produce electrons through the Compton Effect, with electrons further used to generate Cerenkov radiation. The efficiency of the conversion from gamma-ray to optical radiation depends on (Pruett et al., 1984): the target material (the best one proved to be beryllium), its thickness, the distance between the target and the optical fiber, the type of optical fiber, the angle between the gamma-ray trajectory (as a collimating scheme was employed) and the optical fiber axis, the energy of the gamma-ray (investigations were done with a ^{60}Co source - the 1.17 and 1.33 MeV lines, and a ^{24}Na with its 1.38 and 2.76 MeV lines).

Optical materials, as per se or included in optical fibers, exhibit radioluminescence (radiation induced luminescence – RIL): visible radiation is emitted as long as the materials are exposed to ionizing ration. This signal is generally superposed over the Cerenkov spectrum, but can be distinguished by its narrow band emission spectra. The radioluminescent peaks observed, function of the material investigated, its dopants and the type of the irradiation, are located from UV - 185 nm, 250 nm, 285 nm, to visible - 420 nm, 450 nm, 696 nm (Toh et al., 2002; Treadaway et al., 1975; Yoshida et al., 2007). The superposition of the two spectra makes the discrimination of the luminescence signal difficult, and so, several methods to improve the S/N in detecting the radioluminescent signal against the Cerenkov radiation were developed: the gated detection of the luminescence (Justus, 2006; Plazas, 2005), different time frames or the different angular distribution of the two signals are used to detect the RIL (Akchurin et al., 2005); the subtraction method by employing a dummy optical fiber to record the Cerenkov radiation separately; an optical filter or a set-up based on a spectrometer (Archambault, 2005; Lee et al., 2007b; Jang et al., 2010a; Jang et al., 2011a).

An alternative to the generation of radioluminescence in optical materials is represented by the use of commercially available scintillating optical fibers as those delivered by Saint-Gobain Crystals and Detectors – France (Saint-Gobain, 2005) - core material: Polystyrene, core refractive index: 1.6, cladding material: acrylic, cladding refractive index: 1.49, NA: 0.58, scintillation efficiency: 2.4%, trapping efficiency: 3.4%, by Kuraray Co. (Japan) or specially designed ones, as Ce doped optical fibers (Chiodini et al., 2009).

Additional effects produced in silica optical fibers under irradiation refer to:

1. the change of the silica density observed either during gamma irradiation tests on distributed optical fiber sensors based on Brillouin scattering: a non-linear increase with the total dose of the measured frequency and Brillouin FWHM line width (Alasia et al., 2006), or under soft X-ray or reactor irradiation (Primak et al., 1964);

2. the modification of the refractive index of silica subjected to gamma-ray (Fernandez Fernandez et al., 2003).

In concluding this section on silica optical fibers, it is worth mentioning the behaviour of fiber Bragg gratings (FBGs) manufactured in silica optical fibers as they are exposed to irradiation, considering some authors suggestion to use such sensors as radiation detectors (Krebber et al., 2006). Several tests were run to evaluate the FBGs radiation hardness under neutron and gamma-ray exposure, in order to assess their possible use for temperature or stress monitoring in nuclear environments. Various types of FBGs as it concerns the optical fiber material, and the fabrication technology were used: Ge-doped (10 mol % GeO_2, Accutether AT120) with and without H_2 or D_2 loading (Gusarov et al., 2002; Gusarov et al., 2008a), B/Ge co-doped photo sensitive fiber (Maier, 2005), low-P content and N-doped optical fibers (Gusarov et al., 1999a), grating written by classical UV methods or using fs UV laser radiation (Gusarov et al., 2010). Gamma-ray pre-irradiation treatment or UV exposure of the sample before the nuclear reactor in core irradiation was also carried out in order to increase the immunity to irradiation effects (Fernandez Fernandez et al., 2001; Gusarov et al., 2008a). The Bragg gratings were investigated either in transmission or reflection, the parameters tested under irradiation were: the grating peak wavelength shift, the device transmission/reflectivity, the Full-Width Half-Maximum (FWHM), the temperature sensitivity of the device. Studies revealed that the FBGs coating can play an important role in the radiation sensitivity of the sensor (Gusarov et al., 2008). Generally, under gamma irradiation, the FWHM parameter was not affected by the irradiation, while a shift of the peak wavelength was noticed, as being dependent of the total irradiation dose and the dose rate - the wavelength shift is directed towards longer (Caponero et al., 2007; Fernandez Fernandez et al., 2002; Maier et al., 2005a) or shorter wavelengths (Maier et al., 2005), accompanied by a saturating effect and the decrease of the device reflectivity. Depending on the fabrication process and on the optical fiber composition, highly doped Ge optical fibers proved to be less sensitive to gamma irradiation, while hydrogen loaded optical fibers show a higher sensitivity (Gusarov et al., 1999; Gusarov et al., 2002). In some situations, under mixed gamma-neutron irradiation, the FWHM of the FBGs had also increased, suggesting the partial erasing of the written grating (Fernandez Fernandez et al., 2001). The temperature coefficient of the FBG is not affected by gamma irradiation (Gusarov et al., 1999b).

Investigations into the radiation degradation of Poly(methyl methacrylate) (PMMA) (Yoshida & Ichikawa, 1995) show that the effects of ionizing radiation can be divided into two sections, namely main-chain scission (degradation) and crosslinking. In many polymers both processes take place in parallel with one another, however in certain cases the scission predominates the crosslinking, and such polymers are known as degrading polymers. PMMA, shown in Fig. 1, is one such polymer.

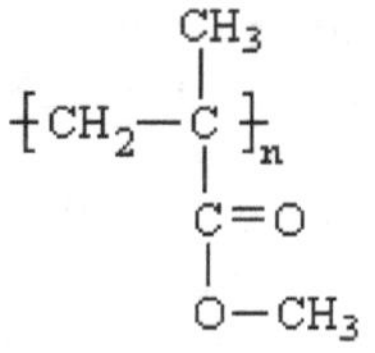

$$\left\{CH_2-\underset{\underset{\underset{O-CH_3}{|}}{\overset{\overset{CH_3}{|}}{\underset{C=O}{|}}}{C}}\right\}_n$$

Fig. 1. PMMA polymer molecule.

Yoshida and Ichiwkawa (Yoshida & Ichikawa, 1995) report that the side-chain is initially affected by the gamma irradiation and the radical formed is a precursor for the main chain scission. When PMMA is irradiated with ionizing radiation, such as gamma radiation, a free radical is generated on the ester side-chain, $-COO\dot{C}H_2$.

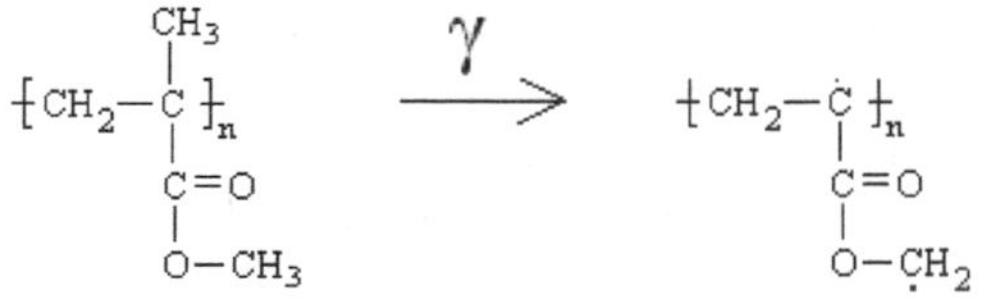

$$\left\{CH_2-\underset{\underset{\underset{O-CH_3}{|}}{\overset{\overset{CH_3}{|}}{\underset{C=O}{|}}}{C}}\right\}_n \quad \xrightarrow{\ \gamma\ } \quad \left\{CH_2-\underset{\underset{\underset{O-\dot{C}H_2}{|}}{\overset{\overset{CH_3}{|}}{\underset{C=O}{|}}}{\dot{C}}}\right\}_n \tag{1}$$

This side chain radical may be generated in a number of ways (Ichikawa, 1995), by direct action of ionizing radiation:

$$-COOCH_3 + \gamma \rightarrow -COO\dot{C}H_2 + H \tag{2}$$

by proton transfer of the side-chain cation:

$$-COOCH_3 + \gamma \rightarrow -COOCH_3^+ + e^-$$
$$-COOCH_3^+ \rightarrow -COO\dot{C}H_2 + H^+ \tag{3}$$

or by hydrogen abstraction:

$$-COOCH_3 + H \rightarrow -COO\dot{C}H_2 + H_2 \tag{4}$$

The side-chain radical remains stable at temperatures below 200 K. Above 210 K, the side-chain radical coverts to a scission type radical due to the detachment of the side chain. This is followed by the β-scission of the main-chain radical and is shown in equations 5 and 6.

$$(5)$$

$$(6)$$

The Beer-Lambert law can be used to determine the radiation dose absorbed. The absorption co-efficient, α, is given by equation 7 (O'Keeffe et al, 2007).

$$\alpha = -\frac{1}{L_0}\log\left\{\frac{P(\lambda)}{P^0(\lambda)}\right\}, \tag{7}$$

where $\dfrac{P(\lambda)}{P^0(\lambda)}$ is the ratio of the spectral radiant power of light transmitted through the irradiated dosimeter to that of the light transmitted in the absence of the dosimeter.
The Beer-Lambert law is used to determine the radiation-induced attenuation. The radiation-induced attenuation is given by equation 8.

$$RIA(dB) = -\frac{10}{L_0}\log\left\{\frac{P_T(\lambda,t)}{P_T^0(\lambda)}\right\}, \tag{8}$$

here L_0 is the irradiated length of fiber, $P_T(\lambda,t)$ is the measured optical power in the irradiated fiber and $P_T^0(\lambda)$ is the optical power of the reference fiber.

3. Environments for use of optical fiber sensing in radiation dosimetry

The possible applications of optical fibers and the optical fiber sensors in radiation monitoring and dosimetry refer to:

1. measurement of the absorbed dose in radiotherapy (Andersen et al., 2002; Jang et al., 2009; Justus et al., 2006) and brachytherapy (Suchowerska et al., 2007);

2. spatial dose distribution in the case of linear electron and proton accelerators for medical treatment (Bartesaghi et al., 2007; Jang et al., 2010a; Lee et al., 2007a);

3. evaluation of beam losses (dose rate, total dose, location) in particle accelerators (Henschel et al., 2004; Intermite et al., 2009; Wulf & Körfer, 2009), beam profiling (Wulf & Körfer, 2009), and the operating conditions of an electron storage ring (Bahrdt et al., 2009; Rüdiger et al., 2008);

4. synchrotron radiation beam profile diagnostics (Byrd et al., 2007 ; Chen et al., 1996);

5. neutron or mixed gamma-ray neutron dosimetry (Bartesaghi et al., 2007a; Jang et al., 2010b);

6. the investigation of isotopic composition of cosmic rays (Connell et al., 1990);

7. radiation dosimetry in computed tomography (Jones & Hintenlang, 2008; Moloney, 2008);

8. distributed radiation dosimetry for beta & gamma rays, and neutrons (Naka et al., 2001);

9. beam profile in the case of free electron lasers (Goettmann et al., 2007) or proton beams (Benoit et al., 2007);

10. remote monitoring of ground water or soil for radioactive contamination (Jones et al., 1993);

11. radiation protection and monitoring of nuclear installations (Magne et al., 2008);

12. monitoring of radioactive waste (Nishiura & Izumi, 2001);

13. reconstruction of the charge particle tracks (Adinolfi et al., 1991; Angelini et al., 1989; Atkinson et al., 1988; Mommaert, 1992; Nakajima et al., 2009 ; Yukihara et al., 2006);

14. the use as transfer detectors for dosimetric calibrations (Tremblay et al., 2010);

15. space dosimetry (Yukihara et al., 2006);

16. tritium detection (Jang et al., 2010c).

In addition, optical fiber-based dosimeters can find in the near future their way in more common environments. One of the main application areas of gamma radiation is in sterilisation, in particular, the sterilisation of medical products. The ability of radiation to kill pathogenic micro-organisms is the basis on which the sterilisation of medical products depends. It is used in a wide variety of products such as hypodermic needles, surgical sutures, blood handling equipment, implant substances and tissues, surgical gloves and utensils, catheters, dental supplies, etc. It is still unknown how exactly the radiation kills micro-organisms, however it is thought to be associated with the damage caused by the radiation to the deoxyribonucleic acid (DNA) of the micro-organism (McLaughlin et al., 1989). The radiation sensitivity of micro-organisms depends on the amount of DNA in the nucleus, but it also depends on a number of other factors, such as the environment they are in. Micro-organisms irradiated in an aerobic environment are more sensitive than those in an anaerobic environment. Also, those in water are more sensitive than those in the dry state. The exact dose of gamma radiation used in the sterilisation of medical products varies for different countries but is usually between 25 kGy and 35 kGy (McLaughlin et al., 1989).

Gamma radiation can be used for a number of different applications within the food industry. Food can be treated to prevent sprouting in onions, garlic and potatoes, to extend the shelf-life of mushrooms, cherries and strawberries, to eradicate insects in grain and fruit, to kill pathogenic microorganisms in fish and meat, to pasteurize dried herbs and spices, and to delay ripening of fruit and vegetables. Radiation can also be used to prevent the spoilage process, which commonly leads to the rotting of food and wastage. This is due to radiation slowing down the physiological, chemical and biochemical changes occurring in the food and killing micro-organisms and insects. Although radiation does not prevent the drying out of fresh food, its application allows the food to be treated in a sealed package, which prevents both the desiccation process and the microbial re-contamination. Food irradiation treatment depends on the ability of radiation to kill cells and alter the enzyme activities of the food. The killing action of radiation, which inhibits cell division, is used to prevent sprouting, reduce the number of viable micro-organisms on the food, prevent the hatching of insect eggs and larvae, and kill or sterilize the insects in the food. A radiation treatment of between 1 kGy to 7 kGy, depending on the product, can significantly reduce the number of viable pathogenic micro-organisms, e.g. *Salmonella*, *Escherichia coli* and *Lysteria*, which can contaminate food causing serious food poisoning (McLaughlin et al., 1989, O'Keeffe & Lewis, 2009).

A number of other applications for radiation are also being investigated, such as in the treatment of sludge, to reduce the amount of bacteria and infectious micro-organisms, before it may be used as a fertiliser. It is also used in the preservation of ancient objects having historic or artistic value by killing the micro-organisms that can destroy the organic material that the objects are made of or have components of (McLaughlin et al., 1989).

4. Experimental set-up and irradiation conditions

Based on the previous considerations, the best assessment of the color centers' dynamics (the equilibrium between color center generation and the recovery process through photobleaching or heating), for different irradiation conditions (dose-rate, temperature, type of radiation) can be achieved during on-line investigations, by optical absorption measurements. Within such an experimental frame, a broadband spectral analysis of the optical transmission characteristics of the optical fiber is recommended in order to catch the complexity of the phenomena involved. In order to achieve this goal we used a set-up composed of combinations of the equipments listed below:

1. broadband light sources, Analytical Instrument Systems DT 1000 CE stabilized deuterium-tungsten lamp for the UV-visible range or a StellarNet SL1 Tungsten Halogen lamp for spectral measurements in the visible-IR domain;

2. a scientific-grade QE65000, TEC-cooled optical fiber spectrometer from Ocean Optics (1024 x 58 pixels; spectral sensitivity from 200 nm to 1100 nm; 0.065 counts/e-sensitivity; 65% quantum efficiency at 250 nm; 1000:1 S/N ratio; 25000:1 dynamic range/single acquisition; 16 bit resolution; from 8 ms to 15 min. integration time; 3 RMS counts dark noise);

3. a NIR InGaAs TEC-cooled mini spectrometer (1024 pixels; spectral sensitivity from 900 nm to 1700 nm; 4000:1 dynamic range, 12 bit resolution; 4000:1 S/N ration; from 1 ms to 30 s integration time) from StellarNet;

a. NIRX-SR InGaAs, 2 TEC stages, mini spectrometer (1024 pixels, spectral sensitivity from 1600 nm to 2300 nm; 4000:1 dynamic range, 16 bit resolution; from 1 ms to 200 ms integration time) from StellarNet;

b.   Avantes optical fiber multiplexers for UV-visible of visible-IR (2 x 8 channels; > 60% optical throughput; > 99% optical repeatability; < 60 ms switching time between adjacent positions).

Several versions of the basic set-up (Sporea & Sporea, 2005) were developed to accommodate various experiments. The overall system is integrated under the National Instruments graphical programming environment and one such experimental implementation is illustrated in Fig. 2 *a*. The experimental instrumentation was used for on-line investigations related to the irradiation effects on silica optical fibers (mostly in the UV range where numerous color centers are developed) and sapphire optical fibers (Sporea & Sporea, 2007; Sporea et al., 2010a; Sporea et al., 2010b). By connecting only one end of the tested optical fiber to different spectrometers, according to the case with or without the use of a multiplexer, the radioluminescent signal was acquired during different types of irradiations. As the temperature plays an important role in recovery of the radiation induced optical absorption some on-line experiments were run by simultaneously irradiating and heating the optical fibers. For gamma-ray irradiation in a cylindrical geometry (the optical fiber, 2 to 5 m long, is coiled and the gamma source is placed in the middle of the circle formed by the fiber) a two sections set-up was used: on the lower level the investigated optical fiber is kept at room temperature, while the optical fiber placed on the higher stage can be electrically heated during the irradiation. A sketch of this mechanical design was presented in Fig. 1 of a previous paper (Sporea et al., 2010a). In that case, the two stages are thermally isolated. For the situation when optical fibers are irradiated by charge particles (electrons or protons) or neutrons, an electrically controlled heater was developed, to rise only locally the temperature of one of the optical fibers, as two short (10 to 15 mm long), similar samples are simultaneously irradiated (Fig. 2 *b*). In this case, each sample is placed into a separate quartz tube, one of these tubes being thermally isolated and coated on the inner side with a heat reflecting layer. In this tube, along with the optical fiber a thermally resistant wire is located parallel to the fiber. By applying a current to this wire the thermally isolated tube is heated. A thermocouple was used to monitor and record in real time the temperature inside the tube. In this way, the optical absorption of both samples is registered in real time, with one of the optical fibers also being subjected to temperature stress.

The set-ups previously detailed were utilized to evaluate the quality of optical fibers suitable for intrinsic or extrinsic radiation monitors (based on optical absorption or radioluminescence) operating at atmospheric pressure. In evaluating materials for alpha particles dosimetry another set-up was designed (Fig. 2 *c*), where the radioluminescence signal is picked-up through a multimode, high UV responsivity optical fiber (Sporea et al., 2010b).

The general conditions for gamma irradiation were described in another paper (Sporea et al., 2010a). In the case of the beta ray irradiation, run at the Linear Accelerator Laboratory of the National Institute for Laser, Plasma and Radiation Physics, the following operating conditions apply: mean electron energy: 6 MeV, the electron beam current: 1 µA, the pulse repetition rate - 100 Hz; the pulse duration - 3,5 µs; the beam diameter - 10 cm; the spot uniformity - +/- 5 %. Proton irradiation was performed at the Tandem irradiation facility of the "Horia Hulubei" National Institute of R&D for Physics and Nuclear Engineering–IFIN-HH, with a beam current of 1 nA, mean dose rate of 200 Gy/s, a total dose of 2.8 MGy, the proton energy of 14 MeV, and the beam diameter of about 3 mm. In some cases, off-line measurements were carried out in order to evaluate the optical absorption recovery either at room temperature or after heating with a laboratory programmable Memmert oven, controlled over a RS-232 serial interface.

 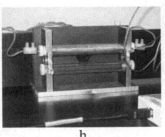 

a   b   c

Fig. 2. Set-ups for the on-line evaluation of irradiation induced optical absorption and radioluminescence: a – the data acquisition and processing system running under LabLIEW graphical programming environment; b – detail of the device used for temperature control of the optical fibers during the irradiation; c – detail of the multimode optical fiber acting as a guiding path of radioluminescence signal generated under alpha particles irradiation (Sporea et al., 2010b).

The two teams jointly investigated, under X-ray irradiation, the characteristics of several extrinsic optical fiber sensors developed at the University of Limerick. Tests in Ireland were done with the ORBITA microfocus X-ray inspection system, while the investigations in Romania were performed with the X-ray micro-tomographic unit developed by Dr. Ion Tiseanu at the National Institute for Laser, Plasma and Radiation Physics. The ORBITA equipment operates within the maximum limits of 0-160 kV, 0- 500 µA. In Romania the sensors were tested with a focused, quasi-monochromatic energy of 17.5 KeV (McCarthy et al., 2011).

5. Optical fiber sensors classification and design

As mentioned in the "Introduction", the use of optical fibers in sensor applications has many advantages over conventional sensors, such as electrochemical and semiconductor sensors. Optical fibers are made of a dielectric material and as such are chemically inert. This makes them very suitable in chemical sensing or in chemically harsh environments. They also provide immunity from electromagnetic interferences and their high electrical isolation makes them suitable for use in lightning protection, high-voltage and medical applications. Optical fibers are also capable of withstanding high temperatures, up to 400 60°C (CeramOptec). These characteristics allow optical fiber sensors to be used in environments not permitted by electrical sensors. Through the use of optical fibers it is possible to have great distances to the measuring point allowing for remote sensing in hostile environments, such as in high-radiation-level areas in the vicinity of a nuclear reactor. The possibility of small, simple interfaces along with lightweight fiber technology results in very appealing sensors. Optical fiber sensors have also been shown to have high sensitivity, large dynamic range along with high resolutions (O'Keeffe, 2008).

The use of plastic optical fibers within the sensor industry is a relatively new technology when compared to glass optical fibers. Plastic optical fibers have been shown to exhibit good transmission in the visible region of the electromagnetic spectrum. Due to the large fiber cross-section of plastic optical fibers, generally a 1 mm core, connecting to the light source and detector is non-problematic. This means that, in contrast to glass optical fibers, no expensive precision components are required for centring the fibers. The large fiber core diameter also means that more light can be transmitted and subsequently incident on the detector, resulting in a higher sensitivity of the sensor. Minor contamination, e.g. dust on the fiber end face, does not result in the complete failure of the sensor system due to its large core diameter. Consequently, fibers can be connected on site in industrial environments with relative ease and without affecting the system. PMMA is also easy to cut, grind and melt and so an uncomplicated process, requiring relatively little time, for processing the end faces is necessary to achieve a clean and smooth surface. Plastic optical fibers are also considered easier to handle when compared to glass optical fibers. Glass fibers tend to break when bent around a small radius, which does not occur with plastic fibers. These fibers are extremely low in cost when compared with glass optical fibers. The properties of PMMA plastic optical fibers result in relatively economical connectors for the system, which further contribute towards a low cost solution (Weinert, 1999).

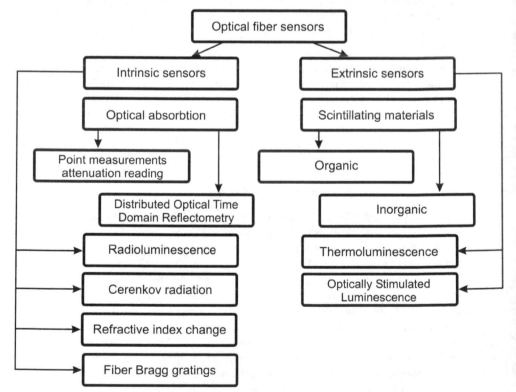

Fig. 3. Classification of the optical fiber sensors for radiation detection according to the operating principle.

Optical fiber sensors can be characterised under two main types: *intrinsic sensors*, where the interaction occurs within the optical fiber itself and *extrinsic sensors*, where the optical fibers are used to guide the light to and from the region where the light interacts with the measurand. Fig. 3 synthesizes the classification of optical fiber-based radiation sensors. Evanescent wave sensors function by causing the light guided in the fiber to couple with the variable to be measured via the evanescent field and have features of both intrinsic and extrinsic devices. It is possible to further classify optical fiber sensors according to the type of modulation to be used (Lopez-Higuera, 1998):

a.   Amplitude or intensity sensors detect the amount of light that is a function of the influencing environment. Their ability to use either incoherent or coherent light sources together with simple optical components makes for a low-cost system.

b.   Phase or interferometric sensors detect a modulation in phase caused by the measurand. This type of source requires coherent light sources making them a relatively costly system. However, they can provide a very high level of sensitivity.

c.   Polarimetric sensors detect a modulation in the polarisation of the light due to the variable to be measured.

d.   Spectroscopic sensors detect a modulation in the spectrum of optical radiation due to the influencing variable.

Radiation dosimetry is fundamental to the wide range of radiation processes and as such is the focus of several research endeavours. A number of physical and chemical sensors, which can be subdivided into liquid, solid and gaseous systems, are available to measure ionising radiation and researchers are continually looking for ways to improve the systems, be it increasing the sensitivity, providing real-time measurements or significantly reducing the costs. All these factors are important for providing the optimum radiation dosimeter. In selecting an optical fiber sensor for radiation measurement (dose rate, total dose, point sensor, 1D or 2D sensors) several parameters have to be considered (O'Keeffe et al, 2008):

a.   **Material** sensitivity: The ability for radiation to interact with the material is the primary characteristic for any radiation dosimeter. This interaction varies greatly among the different types of radiation dosimeters. The materials utilised in gamma dosimetry must have a high sensitivity to gamma radiation within the dose ranges required for the specific application. Especially for in-vivo medical measurements the sensor has to be sensitive to small dose rate/ total doses, comparable to those involved in radiotherapy.

b.   Linearity: The dependency of the sensor out-put signal as function of the input dose/ dose rate. Some applications require a good linearity (no signal saturation) over several decades.

c.   The residual signal: It is required that the output signal after its reading (for the case of re-usable sensors such as those based on thermoluminescence or Optically Stimulated Luminescence) to be as small as possible, no cumulative signal from irradiation to irradiation can be accepted or at least this un-read signal has to be reproducible.

d.   Post-irradiation fading: The material used as a radiation dosimeter must also show minimum post-irradiation fading. This post-irradiation fading is due to the repair of the physical damage caused to the material during irradiation. Information regarding the irradiation dose is lost if the dosimeter material begins to recover immediately after irradiation.

e.   After glow or phosphorescence: in the case of sensors based on radiation induced luminescence a "tail" of the luminescence signal can impinge on the temporal response of the sensor.

f.   Time dependence: Many materials also exhibit a time dependence of specific absorbance. This means that it is often required to wait up to 3 h before obtaining an accurate reading. For real time medical applications it is usually important that the dosimeter is fast to measure.

g.   Stability and reliability: Stability and reliability of the dosimeter, as with any sensor, are also important. In order to ensure that this can be achieved the dosimeter must be immune to a number of environmental conditions. The effect of humidity and temperature on the dosimeter must be investigated and accounted for. The influence of dose rate on the dosimeter material should also be considered. Immunity to other disturbances, such as those found in electromagnetically harsh environments, is also advantageous for gamma dosimeters as they are often employed in such conditions.

h.   Ease of use: For medical applications it is important that the dosimeters are easy to use. For this to be achieved, the system must be easily installed in the area of application without the risk of affecting the measurements. Maintenance should also be minimal. The readout from the dosimeter should be clear and easy to understand. It is also important that the system is low in cost. The availability of clear, easy to implement calibration procedures is an additional advantage.

5.1 Intrinsic optical fiber sensors

Radiation dosimetry based on RIA in gradient index P-doped optical fibers was investigated for ^{60}Co gamma-ray irradiation (dose rates from 0.01 mGy/s to 1.9 Gy/s) and pulsed electrons (total doses of 30 Gy, 100 Gy and 1 kGy). The measurements were carried out at several fixed wavelengths (670 nm, 850 run, 1300 run, and 1550 nm) or spectrally with a spectrum analyser, over the 500 nm to 1700 nm spectral range. In order to be a reliable radiation detector these optical fibers were checked to exhibit high reproducibility, good linearity (over six decades), reduced dose rate dependence (independent over five decades), low fading (a drop of the irradiation induced attenuation of 10 % after two hours and of 14 % after five hours) and small temperature dependence (above 70 ^{0}C the annealing process is significant) and sensitivity to photobleaching (at 10 µW and respectively 10 nW optical power of the laser used to measure the optical absorption no noticeable decrease of the attenuation was observed) (Henschel et al., 1992).

The irradiation induced absorption in doped optical fibers (GeO$_2$ or P$_2$O$_5$, from 12 to 16 mol/%, in the core), under gamma-ray irradiation at low dose rate (1, 0.1 and 0.01 Gy/h, up to a total dose of 1 Gy) was done to evaluate the possible use of such optical fibers as radiation dosimeters. The absorption was monitored during the irradiation (^{60}Co-gamma radiation source of energy 1.25 MeV and Cs-137 radiation source of energy 0.662 MeV) over the wavelength interval 350 nm - 1100 nm (Paul et al., 2009).

As compared to the published data, our investigations focused mostly on the estimation of irradiation effects on special optical fibers, by checking their response in the UV-visible spectral range. We worked with optical fibers such as: solarization resistant; UV-enhanced response; H$_2$-loaded; or operating in the deep-UV, having a core diameter from 200 µm to 1 mm. We evaluated, using the set-up described under headline 4, optical fibers with different jacket materials: Tefzel, Polyimide or Al (Sporea et al., 2010b). Tests were run under gamma or beta rays, and neutron irradiations. Fig. 4 illustrates some of our results recorded over the

UV-visible spectral range (from 200 nm to 950 nm) for optical fibers irradiated at room temperature.

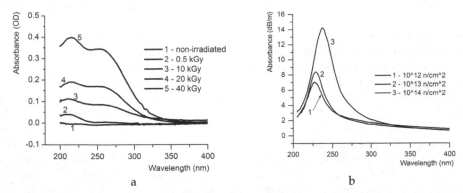

a   b

Fig. 4. The changes of the UV optical absorption for: a - 600 μm core diameter, 12 cm long, UV-enhanced, Polyimide jacket optical fiber, irradiated by beta radiation; b - 400 μm core diameter, 10 cm long, solarization resistant, Tefzel jacket optical fiber, subjected to neutron irradiation at three values of the neutron fluences (Sporea et al., 2010b).

The annealing can play an important role in the recovering process of the radiation induced optical absorption. For this reason we run some on-line experiments to evaluate the impact of the ambient temperature during the irradiation process (Fig. 5). The graphs indicate the modifications of the optical attenuation during on-line measurements for a gamma-ray irradiation (dose rate of 36 Gy/min) 400 μm core diameter, UV-enhanced response, H_2-loaded, solarization resistant optical fiber. The difference in the peak maxima results from the coupling losses along the connecting fibers. The change of the optical attenuation is higher for the sample subjected to heating (240 °C) during the irradiation. Such behaviour can help to select appropriate optical fibers and irradiation conditions in relation to radiation monitoring.

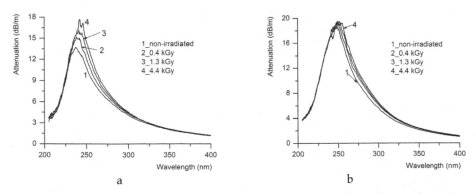

a   b

Fig. 5. The influence on the irradiation induced absorption for a 12 cm long, solarization resistant optical fiber: a – heating to 240 °C is applied to the optical fiber during the irradiation; b – the irradiation is performed at room temperature (Sporea et al., 2010b).

Experimental results (Fig. 4) as well as the corresponding decomposition of the spectral characteristics into Gaussian components (Sporea et al., 2010a) indicate the development of various color centers as the optical fiber is irradiated. In order to estimate the possibility to use UV optical fibers in radiation dosimetry, we measured the change of the optical absorption at two specific wavelengths (Fig. 6 and 7), for different optical fibers, at room temperature, under gamma and beta rays irradiation (Sporea et al., 2010b).

Two wavelengths were monitored in order to optimize the optical fiber as it concerns the radiation sensitivity and linearity of its response with the total dose. The two samples in Fig. 6 a exhibit a very good linearity with the total dose if the color center developed at $\lambda = 240$ nm is considered. A higher sensitivity can be achieved (in the detriment of the dynamic range) with the same optical fibers, if the measurements are done for $\lambda = 215$ nm. The second set of optical fibers (Fig. 6 b), being designed for UV applications, are by far less sensitive to gamma irradiation, but they have also a good linear response with the total dose. The results in Fig. 6 indicate a good dynamic range for the investigated optical fibers. Some of the optical fibers tested under gamma-rays were also evaluated under electron irradiation. They showed to have low radiation sensitivity, and a poorer linearity under beta radiation (Fig. 7). The reading at $\lambda = 280$ nm for the deep-UV optical fiber designed for laser beam delivery remains almost unchanged with the total dose. The color center produced at $\lambda = 215$ nm for the same fiber shows a higher sensitivity and an acceptable linearity up to 40 kGy.

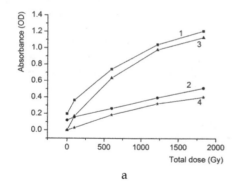
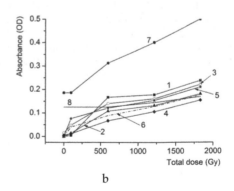

a   b

Fig. 6. The optical absorbance as function of the total dose, gamma-ray irradiation (700 Gy/h) for: a – 1 mm core diameter, multimode silica optical fiber, at $\lambda = 215$ nm (1) and $\lambda = 240$ nm (2); 400 µm core diameter, solarization resistant, Tefzel jacket optical fiber, at $\lambda = 215$ nm (3) and $\lambda = 240$ nm (4); b – 600 µm core diameter, UV-enhanced, Polyimide jacket optical fiber, at $\lambda = 215$ nm (1) and $\lambda = 240$ nm (2); 600 µm core diameter, H_2-loaded, Polyimide jacket optical fiber, $\lambda = 215$ nm (3) and $\lambda = 240$ nm (4); 600 µm core diameter, deep UV-enhanced, Polyimide jacket optical fiber, at $\lambda = 215$ nm (5) and $\lambda = 240$ nm (6); 400 µm core diameter, deep UV improved for laser delivery optical fiber, at $\lambda = 215$ nm (7) and $\lambda = 240$ nm (8) (Sporea et al., 2010b).

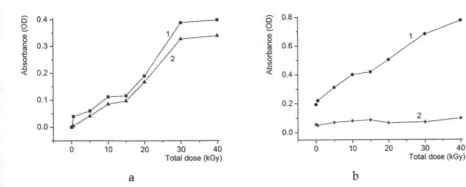

a   b

Fig. 7. The optical absorbance as function of the total dose for beta-ray irradiation (6.1 kGy/min) for: a - 600 μm core diameter, deep UV-enhanced, Polyimide jacket optical fiber, at λ = 215 nm (1) and λ = 280 nm (2); b - 400 μm core diameter, deep UV improved for laser delivery optical fiber, at λ = 215 nm (1) and λ = 280 nm (2) (Sporea et al., 2010b).

The RIA induced by X-ray irradiation in multimode P-doped optical fibers (7 wt.% phosphorus concentration in the optical fiber core) was investigated as preliminary studies on the use of such optical fibers to evaluate the total dose (Girard et al., 2011). Measurements were carried out for a total dose up to 3 kGy, at different temperatures (5 °C; 15 °C; 25 °C, 35 °C and 50 °C), for dose rates of 1 Gy/s; 10 Gy/s and 50 Gy/ s. The optical absorption was monitored over the 350 nm – 900 nm spectral range. No bleaching of the irradiated optical fibers was noticed and the temperature change under the investigation limits plays a minor role. The highest sensitivity for the irradiation induced absorption was observed in the UV part of the spectrum.

The Irish team explored the possible use of change in the optical absorption in POF under irradiation. Fig. 8 shows the transmission spectrum before the fiber was irradiated and after it was irradiated to 3 kGy of gamma radiation. It is immediately clear that, after exposure to gamma radiation, the transmission is attenuated over the entire visible spectrum. The amount of attenuation depends on the wavelength, with lower wavelengths having a significantly higher attenuation than higher wavelengths (O'Keeffe et al., 2007; O'Keeffe & Lewis, 2009; O'Keeffe et al., 2009a).

The possible use of POFs as radiation monitors is exemplified in reference to Fig. 9, when the peak wavelengths were selected for further analysis. It can be seen that as the wavelengths increase, the rate at which the intensity decreases is reduced. 525 nm shows a rapid decrease in intensity and becomes fully attenuated rapidly. In contrast to this, at 650 nm the intensity, while already low to begin with due to the intrinsically high attenuation in PMMA in this region, decreases at a much slower rate and does not become completely attenuated over the course of the experiment.

The sensitivity of the PMMA fiber to ionizing radiation is directly related to the wavelength and thus by selecting the correct wavelength the fiber can be used to monitor over a wide dose range, selecting 525 nm for low doses with high sensitivity and selecting 650 nm for high doses with lower sensitivity.

Radiation-induced attenuation (RIA) calculations, based on equation 8, were performed on the individual wavelengths to determine the exact sensitivity of the fiber to gamma

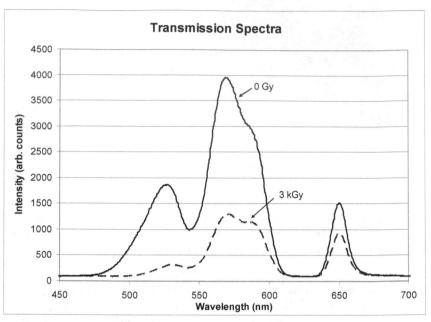

Fig. 8. The transmission of a POF spectra before irradiation and after 3 kGy (O'Keeffe et al, 2009a).

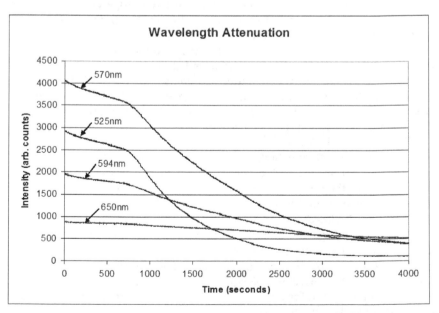

Fig. 9. Intensity of the peak wavelengths over time, as POFs are exposed to gamma radiation (O'Keeffe et al, 2009a).

radiation and the results are presented in Fig. 10 *a*. The wavelength dependence on the RIA sensitivity is evident as the rate of attenuation decreases for increasing wavelength. All the wavelengths exhibit identical characteristics in the RIA with increasing dose. Initially, up to 1 kGy, the RIA increases slowly, followed by significantly higher rate of increase. After this the fiber reaches saturation, whereby the intensity has become completely at that wavelength. The saturation can be seen in Fig. 10 *b*, and indicates the dose range over which the individual wavelengths can monitor. As the wavelength increases the upper-detection limit for that individual wavelength also increases, 4 kGy at 525 nm and 45 kGy at 650 nm (O'Keeffe et al., 2007; O'Keeffe & Lewis, 2009; O'Keeffe et al., 2009a).

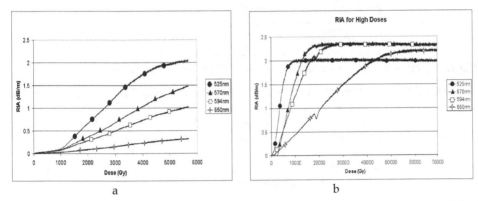

a   b

Fig. 10. Radiation-induced attenuation for gamma irradiation; a – total dose up to 5.7 kGy; b – total dose up to 70 kGy (O'Keeffe et al, 2009).

Fig. 11 shows the radiation-induced attenuation at lower radiation doses up to 50 Gy for the two lower peak wavelengths, 525 nm and 570 nm. The determination of lower detection limit of the fiber was limited by the experimental set-up and the high dose rate of the gamma source used, however it can be seen that there is a distinct and immediate increase in the radiation-induced attenuation at these wavelengths. An averaging filter is applied to the data and indicates that beyond 30 Gy there is a steady, quantifiable increase in the radiation-induced attenuation (O'Keeffe et al., 2007; O'Keeffe & Lewis, 2009; O'Keeffe et al., 2009a).

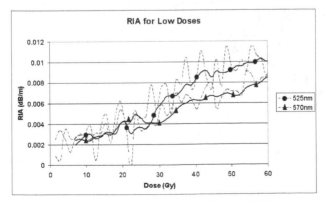

Fig. 11. RIA for low doses of gamma radiation (O'Keeffe et al, 2009a).

Table 1 gives a summary of the sensitivity and dose range of the PMMA fiber for individual wavelengths. It shows that the sensitivity decreases with increasing wavelengths, from 0.6 dBm-1/kGy at 525 nm to 0.06 dBm-1/kGy at 650 nm. The dose range is also dependant on the wavelength as those wavelengths with high sensitivity becoming completely attenuated relatively quickly compared with the higher wavelengths.

Wavelength (nm)	Sensitivity (dBm-1/kGy)	Dosimetry range (kGy)
525	0. 6	0.03 - 4
570	0. 3	0.03 - 10
594	0. 2	0.5 - 12
650	0.06	1 - 45

Table 1. Summary of wavelength dependant sensitivity and dose range.

The radiation induced attenuation of PMMA optical fibers exposed to X-ray radiation, using 6 MV photon energy was reported (O'Keeffe et al, 2009b). The transmission spectra were monitored online and were recorded every two seconds for analysis of attenuation changes as the radiation dose was increased. The results of the RIA calculations are shown in Fig. 12 for two peak wavelengths, 565 nm and 594 nm. There is an immediate and distinct increase in the RIA as the PMMA optical fiber is exposed to the X-ray radiation. The initial results indicate a sensitivity of the fibers to radiation of 0.001 dBm-1/Gy.

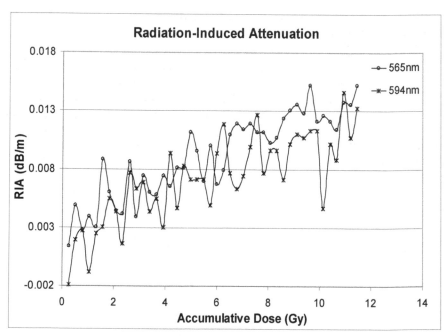

Fig. 12. Radiation Induced Attenuation for two wavelengths, 565nm and 594nm, for POFs under X-ray irradiation (O'Keeffe at al, 2009b).

As the beam losses in particle accelerators have to be checked over long distances a monitoring system, using the principle of Optical Time Domain Reflectrometry (OTDR), was designed for DESY TESLA accelerator (Henschel et al., 2004). The operation of this distributed dosimeter is based on the attenuation induced in optical fibers as they are exposed to radiation associated with machine operation (X-rays, electromagnetic showers). When a short laser pulse is launched into the fiber, some Rayleigh back-scattered optical radiation coming from the fiber sections where color centers are generated by the highly energetic radiation is collected by a detector. From the time of flight and the level of the returned optical radiation pulse a distributed dose mapping can be done, as the pulsed propagation time can be linked to the distance from the laser source of the perturbing non-homogeneities. Within some limits the amount of the detected light can reflect the dose to which the optical fiber segments were exposed. The main limitation of the system is related to the more or less permanent degradation of the optical link transmission during the irradiation. Some solutions to this problem were tested, with not so encouraging results, as for example photobleaching with laser radiation at 830 nm (100 mW CW operation), 670 nm (200 mW/CW), 532 nm (3 W/ CW and 5.5 kW/pulsed).

In the case of P-doped optical fibers, an alternative to extend the operation lifetime of the system can be the thermal annealing (Kuhnhenn et al., 2004). The radiation sensitive optical fiber was subjected to several ^{60}Co irradiation (dose rates between 2.05 to 3.2 Gy/min, total dose of 1000 Gy) - annealing cycles (heating to 400 ^{0}C). A cumulated (after several cycles) residual optical attenuation was observed (about 0.25 dB/m), while the radiation sensitivity of the optical fiber changes very little.

Among the advantages, at least theoretically, of such a system reside in: on-line, distributed dose monitoring; permanent access to difficult to reach locations in complex installations; the possibility to "tune" the sensitivity of the system for specific dose and dose rates, based on the selection of the optical fiber and the interrogation wavelengths used. In order to solve the problem related to the limits imposed by the optical fiber length and the increase in time of the fiber attenuation, an alternative set-up was suggested when the same OTDR units interrogates the optical link from both ends. Even multiple distributed sensors can be operated by the same control instrument (Henschel et al., 2004). Nevertheless, there are some difficulties as it concerns the system performances (the optical fiber radiation sensitivity, the needed information on the exposed optical fiber length, the saturation of the attenuation with total dose). For location where the dose has to be more strictly monitored, a set of local optical fiber dosimeters were designed. In this approach, the attenuation changes with the irradiation dose were evaluated with a LED (λ = 660 nm, for a higher sensitivity) as an interrogation light source, and a set of optical power meters coupled to the LED though connecting optical fibers and segments of radiation sensitive optical fibers, the last ones located in radiation exposed areas.

Apart from standard silica and plastic optical fibers, microstructured optical fibers (MOF) such as photonic crystal fibers (PCF) and random hole optical fiber (RHOF) were tested under irradiation, to assess their radiation hardness or possible use in radiation dosimetry (Wu et al., 2003). In RHOFs the silica core is surrounded by thousands longitudinal channels, having random positions and dimensions. By coating the holes of a RHOF with a phosphorescent material acting as a converter of X-ray or gamma-ray energy to optical radiation, a dose rate monitor can be built, the optical fiber as a whole acting also as a light guide (Alfeeli et al., 2007).

A sensor based on the detection of the Cerenkov radiation produced in a bunch of radiation resistant, high OH-content optical fibers was used to optimize the injection efficiency of the transfer channel between the synchrotron and the storage ring at DELTA. The set-up makes possible a 2D mapping as the time of flight is used to evaluate the location of the Cerenkov radiation produced along the optical fiber, with a longitudinal spatial resolution of 0.2 m corresponding to a delay of about 1 ns between two Cerenkov flashes, while the spatial positioning of the detecting optical fibers assures the localization along the transversal direction (Rüdiger et al., 2008; Wittenburg, 2008). Higher spatial resolution of the Cerenkov radiation-based sensor can be achieved by employing two parallel positioned optical fibers, one used as reference (it is optimized to generate and guide the maximum amount of Cerenkov radiation) and the other one is structured as a sequence of alternating segments of optical fibers similar to the reference fiber and optical fibers having a much higher attenuation. Both optical fibers are coupled to silicon photomultipliers. In this way, if the segments composing the second optical fiber are short enough, by measuring the attenuated signal as compared to the optical signal picked-up by the reference optical fiber, the location of the generated Cerenkov radiation can be identified (Intermite et al., 2009).

A set of four radiation resistant, high-OH content optical fibers, coupled to photomultiplier tubes to detect the Cerenkov radiation, were installed in equidistant radial arrangement along the undulator section of a free electron laser. The position of a wire scanner crossing the electron beam is correlated with the detected Cerenkov radiation, resulting in beam loss, and so, the profile of the beam can be reconstructed by appropriate data processing of the acquired optical signal (Goettmann et al., 2007).

A distributed radiation dosimeter for beta rays, gamma rays and neutrons was designed by using a plastic optical fiber (POF) subjected to irradiation and by detecting the fluorescence and Cerenkov radiation optical signals at both ends of the optical fiber. The time-of-fly technique applied to pulses detected by photomultiplier tubes combined with time-to-amplitude converters and a multichannel analyzer was used to locate the place where the irradiation occurs (Naka et al., 2001). The operating parameters of the equipment (responsivity, resolution) are function of the energy spectrum of the detected radiation, the optical fiber attenuation and degradation, the emission spectrum of the generated optical radiation.

We explored the possible use of sapphire optical fibers in spectroscopic applications in radiation environments and in radiation dosimetry. 25 cm long 425 µm diameter sapphire optical fibers were measured in relation to the radiation induced attenuation (Fig. 13) and radioluminescence, as they were subjected to gamma-ray, proton and neutron irradiation (Sporea & Sporea, 2007). By analyzing the experimental data we noticed the radiation resistance of these optical fibers and the three peaks spectrum of the RL emission under proton irradiation. As the sapphire fibers proved to be immune to radiation (tests were run for gamma-ray up to 530 kGy total dose, for proton up to 2.8 MGy total dose, and for neutron for fluences up to 3.15×10^{13} n/cm^2) we can think to have these optical fibers as real time dose rate monitors, by measuring the RL signal (Fig. 14). Because a small increase of the optical attenuation occurs for $\lambda = 350$ nm to 550 nm after irradiation, dosimetric evaluation based on RL can be performed more accurately if the amplitudes of both emission peaks ($\lambda = 440$ nm and $\lambda = 690$ nm) are considered to correct the green emission measurements. The noisy signal in UV is due to a poor transmission of sapphire under 350 nm. The set-up described in section 4 has the capability to acquire the time dependence of data at selected wavelengths, hence there is the possibility to monitor in real the radioluminescence associated to different centers and to have information on the beam energy.

Another intrinsic type of optical fiber sensor, suggested to be used in particle accelerators for evaluation of very high doses, is based on a Mach-Zehnder interferometer. One of the interferometer arms represents the reference, while the other arm is exposed to irradiation. This second arm is formed by a radiation sensitive optical fiber exhibiting a refractive index change with the radiation dose, and, by combining the coherent radiation traveling along the two arms, a variation of the refractive index produced by the irradiation can be detected (Henschel et al., 2000).

Exploring the radiation induced changes in the refractive index of silica optical fibers a FBG-based dosimeter was suggested, applicable to high radiation doses (10^3 Gy up to 10^5 Gy with an accuracy of at least 10%). Measurements were carried out, at different dose rates, under gamma ray irradiation, assuring a strict control and measurement of the temperature during the irradiation (Krebber et al., 2006). The sensors can be optimized for radiation dosimetry by a careful selection of the optical fiber type and the operating wavelength. It was demonstrated theoretically that FBGs written in photonic crystal fibers (PCF) can achieve a high sensitivity to radiation (at least one order of magnitude) as compared to FBGs produced in bulk material (Florous et al., 2007).

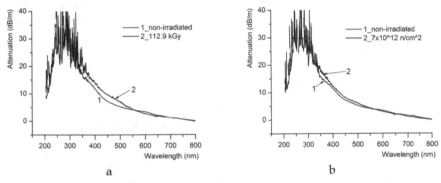

Fig. 13. The changes in the optical attenuation in a 25 cm long, sapphire optical fibers under: a – gamma-ray irradiation; b – neutron irradiation (Sporea & Sporea, 2007).

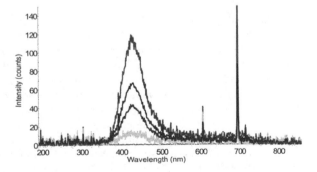

Fig. 14. The radioluminescence emission from a sapphire optical fiber, for proton beam irradiation. The amplitude changes of the RL signal reflects the modifications of the beam current at different irradiation stages (Sporea & Sporea, 2007).

5.2 Extrinsic optical fiber sensors

Point measurements of the absorbed dose in radiotherapy were done by coupling a 10 mm long, 1 mm diameter piece of organic scintillator (Saint Gobain Crystals BCF-20) to a POF acting as light guide for the generated optical radiation. A higher sensitivity of the sensor was obtained by outside coating the scintillating tip by a reflecting TiO_2 layer. The read-out was performed with a photodiode-amplifier system. The device was tested under proton irradiation at a Varian CLINAC 2100C/D accelerator for a 6 MeV energy beam. The amount of the collected optical radiation varied linearly with the dose rate up to 900 cGy/min. The system, acting as a real time instrument, provides within a small volume, high resolution, water equivalent readings, without the need of temperature, humidity or pressure correction (Jang et al., 2009). A similar detecting optical set-up was used to build a 1D array of sensors for high energy proton beam profiling, with scintillating tips spaced 1 cm and respectively 0.5 cm apart, and the light detection with a 10 channel photodiode array (Lee et al., 2007a).

An alternative approach uses a Bicron BC400 scintillating material glued at the end of a PMMA optical fiber and coated with an Al layer, for improved optical radiation collection (Suchowerska et al., 2007). The dimensions of the sensor make possible the limitation of the background optical noise, fluorescence and Cerenkov emission in the optical fiber (less than 0.1 % of the useful signal).

Another optical fiber dosimeter for use in radiotherapy is composed of fused quartz glass doped with Cu^{1+} ions fused to multimode silica optical fiber, which permits the connection to a photomultiplier and a data collection system. Prior to exposure to 6 MV X-rays from a linear accelerator, the radiation detection tip was pre-sensitised under soft X-ray radiation in order to stabilize its responsivity (Justus et al., 2006). To remove the background noise, composed of the native fluorescence in the coupling optical fibers (having a decay time of several ns) and the Cerenkov radiation (emitted in the ps range), and to keep for the dose measurement only the scintillation signal from the Cu^+-doped fused quartz optical fiber (lasting several hundreds ms) a gated data acquisition is performed by temporarily selecting the signal to be processed. The measurements carried out indicated the independence of the system response on the dose rate, energy and dose per pulse. Some limits are associated to the long-term stability, degradation under irradiation, directionality of the response (Tanyi et al., 2011).

An array of eight pieces of scintillating optical fibers (BCF-10 type), interlaced with eight pieces of clear optical fibers (BCF-98), both types having a length of 5 mm, and a diameter of 1 mm, was coupled to 16 photomultiplier tubes through a bunch of connecting optical fibers, to form a low resolution, radiation imaging device. The role of the short, clear optical fibers is to collect the perturbing Cerenkov radiation generated into the detecting system during the exposure to charge particle beams. In this way, the Cerenkov "noise" can be subtracted from the signal delivered by the scintillating optical fibers, improving in this way the detecting S/N. The set-up was used to evaluate the depth dose curves in a PMMA (polylmethylmetracrylate) phantom, over a water depth equivalent of 14 cm for proton, and respectively 3 cm for electron irradiation, the spatial resolution being given by the optical fibers spacing of 1 cm (Bartesaghi et al., 2007a). The tissue equivalent response of the detection scheme is given by the polystyrene core of the scintillating optical fibers. A 2D detection scheme based on the same principle was developed (Bartesaghi et al., 2007b). Attempts were made to embed a clear optical fiber into a boron (B_2O_3) doped scintillating tip in order to detect neutrons.

A 1D dosimeter composed of 25 individual sensors, each one fabricated by optically coupling a 4 mm/ 400 μm diameter piece of fused-quartz glass doped with Cu^{1+} ions to a POF, which make possible the transfer of the luminescence signal from the radiation detector tips to a CCD read-out system, was used for mapping the radiation filed in computed tomography - CT (Jones & Hintenlang, 2008).

Another approach to compensate the detected signal collected from the scintillating optical fiber (BCF-60) for the Cerenkov radiation generated in the guiding optical fibers uses a CCD readout scheme in connection to two dichroic beamsplitters. The guiding optical fibers output both the fluorescence signal from the scintillator and the Cerenkov radiation. The scintillating material was selected so that it emits optical radiation in the 500 nm to 600 nm spectral band, while the Cerenkov radiation is predominant in the UV-blue spectral range. One of the beamsplitters reflects towards the CCD array the optical signals with wavelengths lower than 500 nm (mostly the Cerenkov radiation), while the other directs to the detector array the optical radiation between 500 nm and 600 nm. Using a suitable calibration procedure the amount of Cerenkov radiation is subtracted from the two optical signals detected by the CCD and the irradiation dose can then be computed (Frelin & Hintenlang, 2006).

In a recent implementation, the sensing tip consisted of a piece of BGO (Bismuth Germanate – $Bi_4Ge_3O_{12}$) attached to a POF, while the detection was done using a photomultiplier (Seo et al., 2011). The BGO was selected as being a very efficient gamma-ray to light converter.

An alternative design used a polystyrene-based organic scintillator (BCF-60, Saint-Gobain) coupled to a POF in order to separate the Cerenkov signal from the scintillating signal by wavelength discrimination. The optical signal was detected by a CCD camera (Jang et al., 2010a).

An extrinsic, off-line radiation dosimeter for beta radiation based on thermoluminescence in Ge-doped and Al-doped silica optical fibers was suggested (Yaakob et al., 2011). Samples of such optical fibers were exposed to 6, 9, and 12 MeV electron beams, together with classical TLD-100 detectors in order to evaluate the newly proposed detecting materials sensitivity, responsivity, linearity, stability and reproducibility. The optical signal realised upon annealing (done at 400 °C for 1 h) was read by a Harshaw 4500 TL instrument. Ge-doped optical fibers proved much more sensitive that the Al-doped ones, and exhibit a quite good linearity. The tested optical fibers make repeated usage possible, as no major degradation appeared.

To avoid the optical noise introduced by the Cerenkov radiation in the UV-visible optical spectral range a new type of optical fiber extrinsic dosimeter was proposed. In this approach, a Gd_2O_2S-based scintillator was developed by rare earth ions doping (Pr^{3+}, Yb^{3+} and Nd^{3+}). The scintillating material was optically coupled to a set of optical fibers in order to guide the light to a CCD detector (Takada et al., 1999). In front of the CCD camera the optical signal, guided by the optical fiber bunch, is filtered with an optical high-pass filter. The system has the advantage that the scintillator emits in this case in the IR, hence, the Cerenkov radiation can be cut-down, improving the S/N ratio on the detecting side. To compensate for the radiation induced degradation of the optical fibers transmission the system was complemented with an Optical Time Domain Reflectometer (OTDR). This instrument monitors and localises the increase of the absorption along an optical fiber, of the same type as those used as light guide fibers, which was placed parallel to the connecting optical fibers. In this way, corrections are introduced to compensate for the degradation of the optical links.

Extrinsic optical fiber radiation sensors, proposed in the mid-60s, rely on the Optically Stimulated Luminescence (OSL). The operation of such a detector is based on semiconductor or insulator materials having, in their wide band gap, two additional energy levels. By exposing the material to ionizing radiation, the energy level close to the conduction band is populated, with the number of the trapped electrons being, more or less, proportional to the irradiation dose. At the end of the irradiation process, the material is exposed to an optical radiation at a specific wavelength and the trapped charges are transferred to the second energy level (located also in the band gap, closer to the valence band). By the recombination of these charges, an optical radiation, representing the reading signal of the detector, is emitted (Aznar[a], 2005; Beddar, n.d.; Edmund, 2007). The most common OSL materials used are: $MgS:Ce,Sm$; $SrS:Eu,Sm$; $NaCl:Cu$; $KCl:Eu$; $KBr:In$; $RbBr:Ti$; $BbI:In$; $A-Al2O3$; $CaS:Ce,Sm$; $SrS:Ce,Sm$ (Liu et al., 2008); $KBr:Eu$ (Klein & McKeever, 2008). Such a sensor can exhibit a very low detecting threshold (down to 10 µGy) and a dynamic range over 6 decades (Benoit et al., 2007). Depending on the materials used, the reading optical radiation is in the visible to near-IR spectral range, while the detected optical radiation is mostly in the UV-visible domain. As the two radiations are spectrally separated the same optical fiber can be employed to carry-out both the stimulation signal (from a light source to the ionising radiation detector) and the luminescence one (from the ionising radiation detector towards the optical detector (photomultiplier, CCD, photodiode)), the luminescence material being optically coupled to one end of the signal guiding optical fiber.

An on-line, OSL dosimeter for the evaluation of both dose rate and total dose was developed by optically coupling a solid state dosimeter of carbon doped aluminium oxide ($Al2O3:C$) to an optical fiber (Andersen et al., 2002). During radiation exposure, some electrons are trapped for a short time in lattice defects of the solid state material, and, as they recombine with holes, a radioluminescence (RL) signal is emitted. In the mean time, additional charges continue to be trapped, as previously described, until an optical radiation stimulus, with an appropriate wavelength laser beam (in this case, λ = 532 nm, 20 mW), realise them and, through recombination with holes, a visible radiation is generated through OSL (Aznar, 2005). Both the excitation light and the detected optical signals (RL & OSL) are guided by the same optical fiber from the laser source and towards the detection/ processing unit. A dichroic beamsplitter is used to couple the stimulation laser radiation to the optical fiber and to separate it from the detected signal, during the detection step. The radioluminescence (RL) signal can be correlated to the irradiation dose rate, while the post-irradiation optically stimulated (OSL) signal can be a measure of the total dose, if appropriate calibrations are performed. That is: the RL signal is increasing with the total dose received, and does not depend on the dose rate, acquired dose history or the radiation quantity. If the measurements are done over well defined time intervals, the change of the RL signal corresponding to each time interval is proportional to the dose rate (Andersen et al., 2009). The equipment was tested with good results (linearity, sensitivity to Cerenkov radiation, reproducibility, angular dependence) under gamma ray, X rays and electron beam irradiation. For X-ray related applications the best reproducibility values achieved were 1.9 % (RL) and 2.7 % (OSL) at 1 σ, with a good linearity up to 30 mGy air kerma (Aznar et al., 2005).

According to the previously described detecting scheme (RL measurement during the irradiation and OSL data acquisition after the irradiation was stopped), both the dose rate and the total dose can be calculated from the RL signal. Unfortunately, these results are affected by the Cerenkov radiation. A correction can be made as the total dose computed from the RL signal is evaluated against the total dose obtained form the OSL signal, when no Cerenkov radiation is present. An improvement for this method was proposed when, during the irradiation interval, OSL reading is simultaneously run with the RL signal detection (which in turn is of course accompanied by Cerenkov radiation). In this case, equilibrium has to be established between the filling rate of the trapping levels due to irradiation and their depletion rate produced by optical stimulation. From the acquired data the dose rate and total dose can be derived. The method permits the suppression of the background noise and avoids the saturation of the OSL detector as trapped charges are periodically removed (Gaza et al., 2004). The "dynamic depletion" mechanism was applied also in the case of other OSL materials such as KBr:Eu (Gaza & McKeever, 2006).

The principle of operation discussed above was extended to an array of 15 multiplexed OSL and RL sensors (Magne et al., 2008). In designing such a system several characteristics have to be considered for the validation of the sensor: response linearity, the temperature dependence of the OSL and RL signals, reproducibility of one channel readings and from channel-to-channel, fading (decrease of the read signal occurring between the irradiation and the reading moment).

In designing an OSL sensor aspects related to the dependence of the RL and OSL signals on parameters such as temperature (during the irradiation and optical stimulation) and linear energy transfer – LET (in the case of charge particles: electrons, protons or heavy charged particles – HCP) have to be considered (Edmund, 2007).

A more complex extrinsic radiation dosimeter, capable of evaluating separately gamma-ray and thermal neutrons under a mixed irradiation, was devised (Jang et al., 2010b). It consists of two detectors based on the BCF-20, Saint Gobin scintillating optical fiber coupled through POFs to photomultipliers for the optical radiation detection. In addition, one of the scintillating detectors has a ^{6}Li cap used for the conversion of incoming thermal neutrons energy to alpha particles. The sensor was tested against pure gamma-ray and thermal neutrons irradiation by exposing it, separately, to a ^{60}Co gamma source and ^{252}Cf source of fast neutrons. In the last situation, thermal neutrons were obtained by interposing a 5-cm-thick polyethylene block between the source and the detector. During the gamma irradiation both detectors (with and without the ^{6}Li converter) generated the same amount of optical radiation. In the second experiment, when they were exposed only to a mixed gamma – neutron field, the signal from the detector, which includes also the converting material, provides information on both gamma and neutron irradiation, while the scintillating optical fiber generates optical radiation induced only by the gamma irradiation. By subtracting the second signal from the first one the contribution of thermal neutrons can be deduced. By adding a third, dummy optical fiber, which does not include any scintillating material, the background Cerenkov radiation can be taken into account and the measurement results can be compensated for it (Jang et al., 2011b).

Different types of plastic scintillators are available for gamma, alpha, beta and neutron detection, with some of them having short scintillating constants, or operating under high temperatures (Beddar, Bayramov & Sardarly, 2008).

Detection based on radiation induced luminescence was evaluated for neutron in ZnS-Ag (peak $\lambda = 450$ nm), ZnS-Cu (peak $\lambda = 570$ nm), SrAl$_2$O$_4$-EuDy (peak $\lambda = 500$ nm), and gamma-ray in Al$_2$O$_3$-Cr, EuO$_2$, and GdO$_2$, when such materials were attached at the end of a radiation resistant optical fiber. Such a fiber only has the role in guiding the optical signal to the optical detection part of the sensor (Shikama et al., 2004). Other implementations used LTB:Mn; LTB:Cu and LTB:Cu,Ag,P (Santiago et al., 2009); GaN (Pittet et al., 2009).

Three types of inorganic scintillating material (Gd$_2$O$_2$S:Tb, Y$_3$Al$_5$O$_{12}$:Ce, and CsI:Tl) fixed with an optical epoxy to the end of a bundle of multimode POFs, used to guide the scintillating radiation towards a photomultiplier, were tested as a tritium sensors (Jang et al., 2010c). Radioactive tritium emits beta rays, which incident on the scintillating material produce optical radiation in the 455 nm – 550 nm spectral range. As the energy of the beta radiation used is very low, the detecting material thickness was designed to be 0.1 mm.

When using scintillating materials it is of interest to assess the stability and the degradation of their characteristics under irradiation. Tests were carried out on PWO (Lead Tungstate - $PbWO_4$) LSO (Lu_2-SiO_5:Ce), and LYSO ($Lu_{2(1-x)}Y_{2x}SiO_5$:Ce crystals (Mao et al., 2009). Upon irradiation, such scintillating crystals can exhibit an increase of the optical absorption, a decrease of the sensitivity, post irradiation phosphorescence (afterglow), a change of the spatial uniformity of their response to radiation. Similar investigations were done on other inorganic scintillators such as: BGO - $Bi_4Ge_3O_{12}$, GSO - Gd_2SiO_5, CWO - $CdWO_5$ (Seo et al., 2009).

A comparison between an inorganic scintillating material (Al_2O_3:C) and an organic scitillator (BCF-12) reveals their advantages and drawbacks (Beierholm et al., 2009):

a. phosphorescence following the irradiation is present in Al2O3:C;
b. as charge carriers are trapped during the irradiation, Al2O3:C exhibits a memory effect making it suitable for OSL sensors production;
c. Al2O3:C has a higher sensitivity than the organic scintillator, but this parameter changes increases with the total dose;
d. Al2O3:C is sensitive to temperature;
e. the temporal response constitute an advantage for BCF-12 material (3.2 ns luminescence life time) as compared to 35 ms for Al2O3:C.

Another design of an extrinsic optical fiber radiation detector is based on a phosphorescent tip fixed at the end of a POF. The fiber is prepared by carefully paring back the outer cable insulation of the PMMA fiber using a special tool, exposing a specific length of the PMMA fiber, as shown in Fig. 15 a. The exposed fiber is inserted and secured into a plastic cylinder of known uniform diameter. A plastic sleeve is inserted over the fiber cable to ensure relative centering of the fiber core in the plastic molding housing. Fig. 15 b shows the PMMA fiber inserted into the plastic cylinder. Once secured, the next step is to prepare the scintillating phosphor material and epoxy mix. The epoxy is formed by adding a known quantity of hardener to another known quantity of resin. While the epoxy is in a liquidated state, a known amount of scintillating phosphor material is combined with the liquidised epoxy. The next step is to inject the epoxy/scintillating phosphor mix into the sealed cylinder mold and let the mixture harden. After a set period of time, the scintillating coated fiber is removed from its mold. Fig. 15 c shows the completed fabrication. The fabrication process is described in more detail elsewhere (McCarthy et al., 2010).

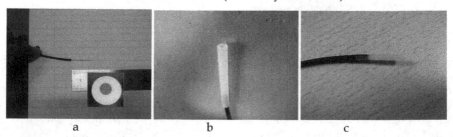

a   b   c

Fig. 15. The preparation sequences of the phosphorous based extrinsic optical fiber radiation sensor (McCarthy et al., 2010)

The resultant fiber optic sensor described above has been tested and found to respond to excitation from low X-ray energies. The fiber optic sensor connects to a computer controlled fluorescence spectrometer which interprets low levels of light received from the sensor upon excitation from an X-ray source. The fiber optic sensor was exposed to 90 keV X-ray energy and the resultant spectra can be seen below in Fig. 16. The peak wavelength response from the sensor was found to be 544 nm. Sensors using different phosphorus were evaluated under X-ray irradiation in Ireland (broadband X-ray source, energy from 20 keV to 140 keV) and Romania (X-ray quasi-monochromatic 17.5 keV source, I = 800 mA, V = 40 kV), and the emission spectra were recorded with the QE65000 scientific grade spectrometer.

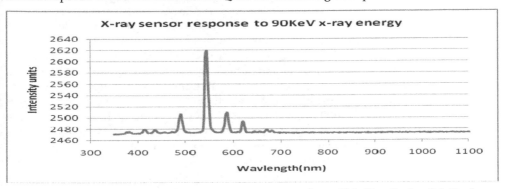

Fig. 16. The spectral response of the X-ray sensor exposed to a 90 keV radiation (McCarthy et al., 2011).

Investigations on the responsivity along the sensor length were carried out by scanning the detector tip along a line parallel to the optical fiber axis (Fig. 17), with a focused (25 µm) X-ray beam (McCarthy et al., 2011).

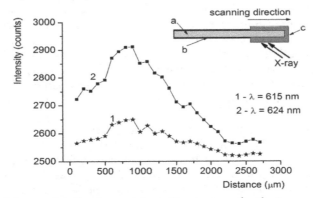

Fig. 17. The spatial responsivity of an extrinsic X-ray sensor for the two peak wavelengths. Insert: a – the POF; b – lightproof jacket; c - phosphorus tip. The focused X-ray beam is scanned along the length of the detecting tip and the profile of the X-ray generated luminescence, as collected by the POSF, is plotted against the distance (McCarthy et al., 2011).

Ge-doped silica optical fibers were investigated as possible thermoluminescence dosimetry (TLD) materials (Hashim et al., 2008). Pieces of 0.8 cm of such optical fibers, carefully measured for their weight (in order to normalize the result to the detector mass) were exposed to X-ray (2 Gy and 10 Gy doses) along with the classical TLD-700 (LiF:Mg,Ti), for comparative evaluation. The signal readout for both materials was done at 300 ^{0}C for 25 s, at a heating cycle rate of 25 ^{0}C/s, with pre-heating at 160 ^{0}C for 10 s. Upon the reading the optical fiber samples were annealed to 400 ^{0}C. The Ge-doped optical fibers have a linear response with the total dose up to 10 Gy. As in the case of other TLD materials, the tested optical fibers exhibit fading (drop of the thermoluminescent signal under lightproof conditions): a decay of 19.7 % after six days storage, as compared to the signal readout after the first 24 h from the irradiation, for the 10 Gy total dose, as compared to the 36.1 % drop for the TLD-700 for the same testing conditions.

Previous studies on commercially available optical fibers (Nokia Cable) used 5 mm long fiber samples irradiated by ^{60}Co gamma radiation at low (from 0.01 Gy to 3 Gy) and high doses (from 4 Gy to 5 kGy). The erase of previously trapped carriers was performed by annealing the samples at temperature from 390 ^{0}C to 490 ^{0}C. The sensor response was compared to a standard TLD-100 dosimeter, as they were subjected to the same irradiation conditions and readings. The peak of the thermoluminescent signal from the optical fiber appeared at about 230 ^{0}C, both for gamma and beta irradiation. Over the entire temperature range, the optical fiber thermally released optical radiation has two peaks, at 400 nm and 575 nm (Espinosa et al., 2006). This investigation did not indicate a significant fading over 15 months of periodic check. The SiO_2 optical fiber presented a higher sensitivity than the TLD-100, by 1.3 times.

In another design, 8 mm long pieces of Ge-doped silica optical fibers were investigated under gamma-ray (^{192}Ir, mean energy 397 keV) and beta (12 MeV energy) irradiation followed by thermoluminescence (TL) reading (Ong et al., 2009). The reported reproducibility is better that 6 %, the sensor having a linear response between 0.2 Gy and 12 Gy for both gamma and beta irradiation. The thermoluminescent signal does not depend on the dose rate and it is angular independent. The fading effect limits the practical use of these optical fibers as thermoluminescent sensors to two weeks.

The same type of optical fibers (Nokia Cable) was evaluated for their possible utilization in OSL dosimetry, as they were exposed to gamma-rays (Espinosa et al., 2011). This optical fiber has a high reproducibility, a low residual signal after several irradiation-reading cycles and, it is estimated from their fading rate, that they can be employed for about six days after the irradiation.

Post irradiation luminescence and the thermoluminescence signals obtained by off-line investigations, from ^{60}Co gamma irradiated COTS optical fibers, are illustrated in Fig. 18 (Sporea et al., 2010b). Depending on the characteristics of the generated optical signal (linearity, reproducibility, sensitivity, etc.) commercially available optical fibers can be incorporated into on-line/off-line radiation detecting schemes, by monitoring the peak value of the emission at specific wavelengths.

Another approach for the design of an extrinsic optical fiber radiation monitor can be developed around Ge-doped silica glass. Figure 19 illustrates the change of the optical absorption and the RL signal as such a glass is irradiated by alpha particles. The optical absorbance measurements were done off-line, while the RL emission was detected on-line using the system described in section 4. In considering the inclusion of such a glass into a real time dose rate meter the same corrections have to be made on the RL signal to compensate for the decrease of the optical transmission of the glass. Such a detector is a single-use type as far as no recovery of the radiation induced color centers was observed (Sporea et al., 2010b). The optical absorption is linked to the total deposed charge, while the RL signal is a measure of the instant beam current.

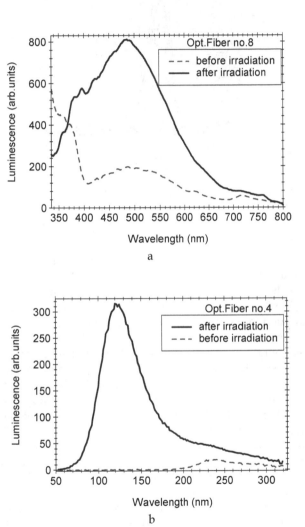

Fig. 18. The luminescence (a) and the (b) thermoluminescence signals after ⁶⁰Co gamma irradiation (total dose 300 Gy, dose rate 158 Gy/h) for: a multimode step-index 400 μm silica core, visible/near-IR optical fiber, with acrylate coating (a); a 600 μm core diameter, deep UV-enhanced, Polyimide jacket optical fiber (b) (Sporea et al., 2010b).

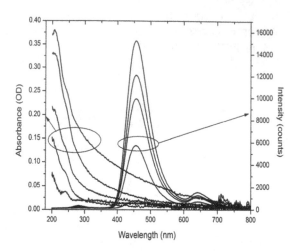

Wavelength (nm)

Fig. 19. The changes induced by alpha particle irradiation in a Ge-doped silica glass: a – incease of the optical absorption (left site curves), for various total deposed charge; b – the radioluminescence signal (right site curves), for different instant values of the beam current (Sporea et al., 2010b)

An optical fiber-based system for on-line dosimetry was designed around an optically transparent thermoluminescent glass material containing ZnS nanocrystals and Cu^{11} ions (Huston et al., 1996). In this approach, the thermoluminescent material is optically coupled to a multimode optical fiber. Nd^{3+} ions were additionally included into the ZnS nanocrystals to increase its absorption for the radiation from a semiconductor laser diode emitting 1 W at 807 nm, which is directed through this optical fiber towards the ionizing radiation (the irradiation was done with a ^{60}Co source) detecting material. The laser radiation is used in this case to heat the thermoluminescent material, which generates an optical readout signal. This signal is picked-up by the same optical fiber and is directed towards a photomultiplier by a dichroic beamsplitter combined with some color filters. The luminescence signal has in this application two TL peaks at 160 °C and 220 °C. The sensor presents a good linearity with the total dose of doses of medical interest (up to 6 Gy).

Earlier, a multiplexed version was proposed in which the output of a laser diode ($\lambda = 840$ nm) coupled to an optical fiber is switched between several "probe" optical fibers. Each such "probe" has fixed at its other end an ionizing radiation detection element. Two such detecting elements were suggested: one formed by a phosphorescent material covered by an absorber and glutted directly to the optical fiber end, or, alternatively, a piece of phosphor, covered by an absorber and mounted to the optical fiber, but isolated with a gap of about 0.1 mm (Jones et al., 1993). As the laser radiation strikes the absorber, its energy is transferred to the phosphor and the thermoluminescent effect appears (at about 400 mW of optical power, reading temperature of 220 °C – 240 °C) if the radiation detector was previously irradiated. The generated optical signal is collected by the same system of optical fiber probes and is directed through a wavelength selective beamsplitter towards a photomultiplier. In the first approach, the optical coupling for the thermally generated signal is a better one, but part of the energy of the laser reading beam is also transferred to the coupling optical fiber. A better sensitivity, from the thermoluminescence point of view, is achieved with the second solution, which has the drawback of a poorer optical coupling for the reading signal. The spectrum of the emitted optical radiation, as well as the temperature of the peak emission, depend on the phosphorus used (in the reported case they are: $CaSO_2:Mn$; $CaF_2:Tm$; $CaF_2:Dy$; $CaSO_4:Dy$). The system proved a good linearity over the 1 mGy to 100 Gy range, under ^{60}Co irradiation. In this application it is worth underlining the opportunity to operate with one reading laser and one detector a multitude of multiplexed miniature dosimeters.

6. Future work

Considering the intrinsic characteristics of optical fiber systems: capabilities to work under strong electromagnetic fields; possibility to carry multiplexed signals (time, wavelength multiplexing); small size and low mass; ability to handle multi-parameter measurements in distributed configuration; possibility to monitor sites far away from the controller, their availability to be incorporated into the monitored structure, numerous groups around the world investigated in the past 20 years the possibility to include optical fiber-based devices into radiation monitoring systems. Most of the implementations relay on radiation induced effects such as: increase of the optical absorption on the optical fiber, generation of optical radiation during the irradiation conditions or post irradiation. The chapter evaluates the main solutions promoted for radiation detectors using the optical fiber as a detecting medium or as a light guide for optical signal transmission. The results obtained up to now are encouraging, but the investigative efforts have to continue as it concerns some basic parameters of these devices: sensitivity, linearity, immunity to dose rate and temperature, dynamic range, response time, etc. Besides the main types of optical fibers mentioned in this review, additional research has to be carried out on novel species of optical waveguides (i.e. microstructured optical fibers, photonic optical fibers, mid-IR optical fibers, new scintillating optical fibers). Tests were also carried out on some plastic optical fibers to evaluate their use as total dose monitors considering the increase of their optical absorption at specific wavelengths.

Further investigations have to be done to check for the reproducibility and the stability of the irradiation induced effects for the proposed optical fibers, as well as the possibility to develop re-usable sensors. The level of the fading effect in the case of sensing materials interrogated by thermoluminescnece has to be verified.

For the radioluminescence detection with sapphire optical fibers and doped silica glasses we intent to proceed to the evaluation of the linearity of the emitted signal as function of the beam parameters (beam current, focusing system current, etc.).

The proposed extrinsic POF sensor has to be improved as it concerns the uniformity of the luminescence signals along the sensing tip, as well as the optical coupling of the luminescence emission to the guiding optical fiber. Additionally, we intend to check its linearity for X-rays irradiation, at higher energies. An extension of the research done on this sensor will be the assessment of beta-ray induced luminescence, in conjunction with the evaluation of effects produced by those irradiations (X-ray and beta-ray) on the optical transmission of the coupling optical fibers.

7. Conclusions

An extended review of optical fiber radiation sensors was introduced, with these type of sensors finding applications in various fields, from nuclear industry to radiotherapy, from space science to particle accelerators. Radiation effects in optical fibers (silica glass, plastic and sapphire) were described along with the classification of the detectors based on intrinsic and extrinsic designs. Examples of authors' experimental results were also included in relation to the tests carried out on silica, plastic and sapphire optical fibers. The evaluation of such optical fibers for the development of radiation monitoring sensors was described.

The chapter emphasis the advantages optical fibers can have in radiation dosimetry: real time measurements (under specific design); small size (able to be used as inside the body or implanted sensors); remote investigations (useful for radiation therapy, particle accelerators, nuclear sites); able to be included into multiplexed or distributed sensor arrays (particle beam diagnostics, monitoring of contaminated sites); susceptible for passive (luminescence, Cerenkov radiation detectors) or active (thermoluminescence, Optically Stimulated Luminescence – OSL) sensing, some of them being water equivalent (an advantage for medical applications).

8. Acknowledgements

The work was done in the frame of the bilateral collaboration existing between the National Institute for Laser, Plasma and Radiation Physics in Bucharest, and the University of Limerick, as part of the COST Action TD1001: *Novel and Reliable Optical Fibre Sensor Systems for Future Security and Safety Applications (OFSeSa)*. The Romanian authors acknowledge the financial support of the Romanian Ministry for Education, Research, Sport and Youth in the frame of the research grant 12084/2008. They also want to thank to their colleagues: Dr. S. Agnello, for providing the Ge-doped silica glass sample, Ms. R. Georgescu and Mr. D. Negut for the gamma irradiation, Dr. C. Oproiu for performing the beta-ray irradiation, Dr. M. Secu for running the luminescence and thermoluminescence measurements, Dr. I. Tiseanu for assisting with X-ray irradiation, Dr. I. Vata for help in doing alpha particles and neutron irradiations. The Irish team wishes to acknowledge the support of the European Commission under the 7th Framework Programme through the 'Marie Curie Re-integration' action of the 'Peoples' Programme, (PERG04-2008-239207).

9. References

Adinolfi, M.; Angelini, C.; Antinori, F.; Beusch, W.; Cardini, A.; Crennell, D.J.; De Vincenzi, M.; Da Vi, C.; Di Paolo, M.; Di Vita, G. ; Duane, A.; Fabre, J.-P.; Flaminio, V.; Frenkel, A.; Gys, T.; Harrison, K.; Lamanna, E. ; Lucchesi, D.; Martelloti, G.; McEwen, J.G.; Morrison, D.R.O.; Penso, G.; Petrera, S.; Roda, C.M.A.; Sciubba, A.; Villalobos-Baillie, O. & Websdale, D.M. (1991). Application of a scintillating-fibre detector to the study of short-lived particle, *Nucl. Instr. Met. Phys. Res.*, Vol. A310, (1991), (485-489)

Akchurin, N.; Atramentov, O.; Carrell, K.; Gümüş, K.Z.; Hauptman, J.; Kim, H.; Paar, H.P.; Penzo, A. & Wigmansa, R. (2005). Separation ofscintillation and Cherenkov light in an optical calorimeter, *Nucl. Instr. Met. Phys. Res.*, Vol. A550, (2005), (185-200), doi:10.1016/j.nima.2005.03.175

Alasia, D.; Fernandez Fernandez, A.; Abrardi1, L.; Brichard, B. & Thévenaz, L. (2006). The effects of gamma-radiation on the properties of Brillouin scattering in standard Ge-doped optical fibres, *Meas. Sci. Technol.*, Vol. 17, (2006), (1091–1094), doi:10.1088/0957-0233/17/5/S25

Alessi, A.; Agnello, S.; Sporea, D.G.; Oproiu, C.; Brichard, B. & Gelardi, F.M. (2010). Formation of optically active oxygen deficient centers in Ge-doped SiO_2 by γ- and β-ray irradiation, *J. Non-Cryst. Solids*, Vol. 356, (2010), (275–280), doi:10.1016/j.jnoncrysol.2009.11.016

Alfeeli, B.; Pickrell, G.; Garland, M.A. & Wang, A. (2007). Behavior of random hole optical fibers under gamma ray irradiation and its potential use in radiation sensing applications, *Sensors*, Vol. 7, (2007), (676-688)

Andersen, C.E.; Aznar, M.C.; Bøtter-Jensen, L.; Bäck, S.Å.J.; Mattsson, S. & Medin, J. (2002), Development of optical fibre luminescence techniques for real time in vivo dosimetry in radiotherapy, *Proceedings of the the International Symposium "Standards and codes of practice in medical radiation dosimetry"*, Vol. 2, pp. 353-360, Vienna, November 2002

Andersen, C.E.; Nielsen, S.K.; Greilich, S.: Helt-Hansen, J.; Lindegaard, J.C. & Tanderup, K. (2009). Characterization of a fiber-coupled Al_2O_3:C luminescence dosimetry system for online *in vivo* dose verification during ^{192}Ir brachytherapy, *Med. Phys.*, Vol. 36 No. 3, (March 2009), (708-718), DOI: 10.1118/1.3063006

Angelini, C.; Beusch, W.; Bloodworth, I.J.; Carney, J.N.; Crennell, D.J.; De Vincenzi, M.; Duane, A.; Fabre, J.P.; Fisher, C.M.; Flaminio, V.; Frenkel, A.; Harrison, K.; Hughes, P.; Kinson, J.B.; Lamanna, E.; Leutz, H.; Martellotti, G.; McEwen, J.G.; Morrison, D.R.O.; Penso, G.; Petrera, S.; Quercigh, E.; Roda, C.; Sciubba, A.; Villalobos-Baillie, O.; Votruba, M.F. & Websdale, D.M. (1989). High-resolution tracking with scintillating fibers, *Nucl. Instr. Methods Phys. Res.*, Vol. A277, (1989), (132-137)

Archambault, L. (2005). *Elaboration d'un dosimeter à fibres scintillantes*, PhD Thesis, Faculté des études supérieures de l'Université Laval

Arvidsson, B.; Dunn, K.; Issever, C.; Huffman, B.T.; Jones, M.; Kierstead, J.; Kuyt, G.; Liu, T.; Povey, A.; Regnier, E.; Weidberg, A.R.; Xiang A. & Yef, J. (2009). The radiation tolerance of specific optical fibres exposed to 650 kGy(Si) of ionizing radiation, *JINST 4 P07010*, (2009), doi:10.1088/1748-0221/4/07/P07010

Atkinson, M.N. ; Crennell, D.J.; Fisher, C.M. ; Hughes, P.T.; Kirkby, J.; Fent, J.; Freund, P.; Osthoff, A.; & Pretzl, K. (1988). A high resolution scintillating fiber (SCIFI) tracking device with CCD readout, *Nucl. Instr. Met. Phys. Res.*, Vol. A263, (1988), (333-342)

Aznar, M.C. (2005), *Real-time in vivo luminescence dosimetry in radiotherapy and mammography using Al_2O_3:C*, PhD Thesis, University of Copenhagen, 2005

Aznar, M.C.; Hemdal, B.; Medin, J.; Marckmann, C.J.; Andersen, C.E.; Bötter-Jensen, L.; Andersson, I & Mattsson, S. (2005). In vivo absorbed dose measurements in mammography using a new real-time luminescence technique, *Brit. J.Radiol.*, Vol. 78, (2005), (328–334), doi: 10.1259/bjr/22554286

Bahrdt, J.; Feikes, J.; Frentrup, W.; Gaupp, A.; v. Hartrott, M.; Scheer, M. & Wüstefeld, G. (2009). Cherenkov fibers for beam diagnostics at the metrology light source, *Proceedings of the 23rd Particle Accelerator Conference*, paper TU5RFP029, Vancouver, May 2009

Bartesaghi[a], G.; Conti, V.; Prest, M.; Mascagna, V.; Scazzi, S.; Cappelletti, P.; Frigerio, M.; Gelosa, S.; Monti, A.; Ostinelli, A.; Mozzanica, A.; Bevilacqua, R.; Giannini, G.; Totaro, P. & Vallazza E. (2007). A real time scintillating fiber dosimeter for gamma and neutron monitoring on radiotherapy accelerators, *Nucl. Instr. Methods Phys. Res.*, A, Vol. 572, (2007), (228-230), doi:10.1016/j.nima.2006.10.323

Bartesaghi[b], G.; Conti, V.; Bolognini, D.; Grigioni, S.; Mascagna, V.; Prest, M.; Scazzi, S.; Mozzanica, A.; Cappelletti, P.; Frigerio, M.; Gelosa, S.; Monti, A.; Ostinelli, A.; Giannini, G. & Vallazza E. (2007). A scintillating fiber dosimeter for radiotherapy, *Nucl. Instr. Methods Phys. Res.*, Vol. A581, (2007), (80-83), doi:10.1016/j.nima.2007.07.032

Bayramov, A.A. & Sardarly, R.M. (2008). Applications of plastic sensors, *Fizika*, Vol. CILD XIV, No. 3, (2008), (149-153)

Beddar, S. (n.d). Scintillation dosimetry: Review, new innovations and applications, http://www.aapm.org/meetings/09SS/documents/32Beddar-PlasticDosimeters.pdf

Beierholm, A.R.; Andersen, C.E.; Lindvold, L.R.; Kjær-Kristoffersenc, F. & Medinc, J., A comparison of BCF-12 organic scintillators and Al_2O_3:C crystals for real-time medical dosimetry, *Radiat. Meas.*, Vol. 43, (2008), (898-903), doi:10.1016/j.radmeas.2007.12.032

Benoit, D.; Vaillé, J-R.; Ravotti, F.; Garcia, P. & Dusseau, L. (2007). Optically Stimulated Luminescence Dosimetry, *1st Workshop on Instrumentation for Charged Particle Therapy*, London, May 2007

Berghmans, F. (2006). Ionizing radiation effects on optical components, *NATO Advanced Study Institute, Optical Waveguide Sensing & Imaging in Medicine, Environment, Security & Defence*, Gatineau, October 2006

Berghmans, F.; Brichard, B.; Fernandez Fernandez, A.; Gusarov, A.; Van Uffelen. M. & Girard, S. (2008). An Introduction to Radiation Effects on Optical Components and Fiber Optic Sensors, *Optical waveguide sensing and imaging*, Bock, W.J.; Gannot, I. & Tanev, S., pp. 127-166, Springer Series B: Physics and Biophysics, Dordrecht, The Netherland

Bisutti, J.; Girard, S. & Baggio, J. (2007). Radiation effects of 14 MeV neutrons on germanosilicate and phosphorus-doped multimode optical fibers. *J. Non-Cryst. Solids*, Vol. 353, (2007), (461-465), doi:10.1016/j.jnoncrysol.2006.10.013

Brichard, B.; Borgermans, P.; Fernandez Fernandez, A.; Lammens, K. & Decréton, M. (2001). Radiation effect in silica optical fiber exposed to intense mixed neutron-gamma radiation field, *IEEE T. Nucl. Sci.*, Vol. 48, No. 6, (December 2001), (2069-2073)

Brichard, B.; Fernandez Fernandez, A.; Ooms, H.; Berghmans, F.; Decréton, M.; Tomashuk, A.; Klyamkin, S.; Zabezhailov, M.; Nikolin, I.; Bogatyrjov, V.; Hodgson, E.; Kakuta, T.; Shikama, T.; Nishitani, T.; Costley, A. & Vayakis, G. (2004). Radiation-hardening techniques of dedicated optical fibres used in plasma diagnostic systems in ITER, *J. Nucl. Mater.*, Vol. 329-333, (2004), (1456-1460), doi:10.1016/j.jnucmat.2004.04.159

Brichard, B. & Fernandez Fernandez, A. (2005). Radiation effects in silica glass optical fibers, In: *Short Course Notebook, New Challenges for Radiation Tolerance Assessment*, pp. 95-138, Cap d'Agde, September 2005

Byrd, J.M.; De Santis, S. & Yin, Y. (2007). Fiberoptics-based instrumentation for storage ring beam diagnostics, *Proceedings of 8th European Workshop on Beam Diagnostics and Instrumentation for Particle Accelerator*, pp. 325-327, Venice, May 2007

Calderón, A.; Martínez-Rivero, C.; Matorras, F.; Rodrigo, T.; Sobrón, M.; Vila, I.; Virto, A.L.; Alberdi, J.; Arce, P.; Barcala, J.M.; Calvo, E.; Ferrando, A.; Josa, M.I.; Luque, J.M.; Molinero, A.; Navarrete, J.; Oller, J.C.; Valdivieso, P.; Yuste, C.; Fenyvesi, A. & Molnár, J. (2006). Effects of γ and neutron irradiation on the optical absorption of pure silica core single-mode optical fibres from Nufern, Nucl. Instr. Met. Phys. Res., Vol. A 565, (2006), (599–602), doi:10.1016/j.nima.2006.05.228

Cannas, M.; Lavinia, V. & Roberto, B. (2008). Time resolved photoluminescence associated with non-bridging oxygen hole centers in irradiated silica, Nucl. Instr. Met. Phys. Res., Vol. B, No. 266, (2008), (2945–2948), doi:10.1016/j.nimb.2008.03.144

Caponero, M.A.; Baccaro, Stefania; Donisi, D.; Fabbri, F. & Pillon, M. (2007). Characterisation of FBG sensors under ionizing radiation for high energy physics and space physics, In: Proceedings of the 10th Conference Astroparticles, Particles and Space Physics, Detectors and Medical Physics Applications, pp. 533-539, Como, October 2007, doi: 10.1142/9789812819093_0092

Cerenkov, P.A., (1958). Radiation of particles moving at a velocity exceeding that of light, and some of the possibilities for their use in experimental physics, Nobel Lecture, December, 1958

Chiodini, N.; Vedda, A.; Fasoli, M.; Moretti F.; Lauria, A.; Cantone, M.-C.; Veronese, I.; Tosi, G.; Brambilla, M.; Cannillo, B.; Mones, E.; Brambilla, G. & Petrovich, M. (2009). Ce doped SiO_2 optical fibers for remote radiation sensing and measurement, Proceeidngs of Fiber Optic Sensors and Applications VI. SPIE Defense and Security, Udd, E.; Du, H.H. & Wang, A., Vol. 7316, doi.org/10.1117/12.818507

Connell, J.J.; Binns, W.R.; Dowkontt, P.F.; Epstein, J.W.; Israel, M.H.; Klann, J.; Webber, W.R. & Kish, J.C. (1990). The scintillating isotope experiment: BEVALAC calibrations of the test models, Nucl. Instr. Methods in Phys. Res., Vol. 294, (1990), (335-350)

Deparis, O.; Griscom, D.L.; Mégret, P.; Decréton, M. & Blondel, M. (1997). Influence of the cladding thickness on the evolution of the NBOHC band in optical fibers exposed to gamma radiations, J. Non-Cryst. Solids, Vol. 216, (1997), (124-128)

Edmund, J.M. (2007). Effects of temperature and ionization density in medical luminescence dosimetry using Al_2O_3:C, PhD Thesis, University of Copenhagen, 2007

Espinosa, G.; Golzarri, J.I.; Bogard, J. & García-Macedo, J. (2006). Commercial optical fiber as TLD material, Radiat. Prot. Dosim., Vol. 119, No. 1-4, (2006), (197-200), doi: 10.1093/rpd/nci564

Espinosa, G. (2011). A study and characterization of the optically stimulated luminescence response of commercial SiO2 optical fiber to gamma radiation, Rev. Mex. Fís.,Vol. 57, No. 1, (2011), (30–33)

Fernandez Fernandez, A.; Berghman, F.; Brichard, B.; Decréton, M.; Gusarov, A.I.; Deparis, O.; Mégret, P.; Blondel, M. & Delchambre, A. (2001). Multiplexed fiber Bragg grating sensors for in-core thermometry in nuclear reactors, Fiber Optic Sensor Technology II, Proc. SPIE, Vol. 4204, pp. 40-49 , March 2001, doi:10.1117/12.417427

Fernandez Fernandez, A.; Brichard, B., Berghmans, F. & Decréton, M. (2002). Dose-rate dependencies in gamma-irradiated fiber Bragg grating filters, IEEE T. Nucl. Sci., Vol. 49, No. 6, (December 2002), (2874-2878), doi: 10.1109/TNS.2002.805985

Fernandez Fernandez, A.; Brichard, A. & Berghmans, F. (2003). In situ measurement of refractive index changes induced by gamma radiation in germanosilicate fibers,

IEEE Photonic Tech. L., Vol. 15, No. 10, (October 2003), (1428-1430), doi: 10.1109/LPT.2003.818247

Florous, N.J.; Saitoh, K.; Murao, T. & Koshiba, M. (2007). Radiation dose enhancement in photonic crystal fiber bragg gratings: towards photo-ionization monitoring of irradiation sources in harsh nuclear power reactors, *Quantum Electronics and Laser Science Conference,* QELS '07, Baltimore, 2007, doi: 10.1109/QELS.2007.4431290

Friebele, E.J. (1991). Correlation of single mode fiber fabrication factors and radiation response, *Naval Research Laboratory _NRL_ Final Report No. NRL/MR/6505_92-6939,* November 1991

Frelin, A.-M.; Fontbonne, J.-M.; Ban, G.; Batalla, A.; Colin, J.; Isambert, A.; Labalme, M.; Leroux, T. & Vela, A. (2006). A new scintillating fiber dosimeter using a single optical fiber and a CCD camera, *IEEE T. Nucl. Sci.,* Vol. 53, No. 2, (June 2006), (1113-1117), doi:10.1109/TNS.2006.874931

Gaza; R.; McKeever, S.W.S.; Akselrod, M.S.; Akselrod, A.; Underwood, T.; Yoder, C.; Andersen, C.E.; Aznar, M.C.; Marckmann, C.J. & Bötter-Jensen, L. (2004). A fiber-dosimetry method based on OSL from Al_2O_3: C for radiotherapy applications, *Radiat. Meas.,* Vol. 38, (2004), (809–812), doi:10.1016/j.radmeas.2003.12.004

Gaza, R. & McKeever, S.W.S. (2006). A real-time, high-resolution optical fiber dosimeter based on optically stimulated luminescence (OSL) of KBr:Eu, for potential use during the radiotherapy of cancer, *Radiat. Prot. Dosim.* (2006), Vol. 120, No. 1–4, (14–19), doi:10.1093/rpd/nci603

Girard[a], S.; Keurinck, J.; Boukenter, A.; Meunier, J.-P.; Ouerdane, Y.; Azaïs, B.; Charre, P. & Vié, P. (2004). Gamma-rays and pulsed X-ray radiation responses of nitrogen-, germanium-doped and pure silica core optical fibers, *Nucl. Instr. Met. Phys. Res.,* Vol. B215, (2004), (187–195), doi:10.1016/j.nimb.2003.08.028

Girard[b], S.; Keurinck, J.; Ouerdane, Y.; Meunier, J.-P. & Boukenter, A. (2004). γ-rays and pulsed X-ray radiation responses of germanosilicate single-mode optical fibers: influence of cladding codopants, *J. Lightwave Technol.,* Vol. 22, No. 8, (2004), (1915 – 1922), doi: 10.1109/JLT.2004.832435

Girard, S.; Vincent, B.; Ouerdane, Y.; Boukenter, A.; Meunier, J.-P. & Boudrioua, A. (2005). Luminescence spectroscopy of point defects in silica-based optical fibers, J. Non-Cryst. Solids, Vol. 351, (2005), (1830–1834), doi:10.1016/j.jnoncrysol.2005.04.043

Girard, S.; Ouerdane, Y.; Boukenter, A. & Meunier, J.-P. (2006). Transient radiation responses of silica-based optical fibers: influence of modified chemical-vapor deposition process parameters, J. Appl. Phys., Vol. 99, (2006), doi: 10.1063/1.2161826

Girard, S.; Ouerdane, Y.; Origlio, G.; Marcandella, C.; Boukenter, A.; Richard, N.; Baggio, J.; Paillet,P.; Cannas, M.; Bisutti, J.; Meunier,J.-P. & Boscaino, R. (2008). Radiation effects on silica-based preforms and optical fibers — I: experimental study with canonical samples, *IEEE T. Nucl. Sci.,* Vol. 55, No. 6, (December 2008), (3473-3482), doi: 10.1109/TNS.2008.2007297

Girard, S.; Marcandella, C.; Origlio, G.; Ouerdane, Y.; Boukenter, A. & Meunier, J.-P. (2009). Radiation-induced defects in fluorine-doped silica-based optical fibers: Influence of a pre-loading with H_2, *J. Non-Cryst. Solids,* Vol. 355, (2009), (1089–1091), doi:10.1016/j.jnoncrysol.2008.11.035

Girard, S. & Marcandella, C. (2010). Transient and Steady State Radiation Responses of Solarization-Resistant Optical Fibers, *IEEE T. Nucl. Sci.*, Vol. 57, No. 4, (2049 – 2055), doi: 10.1109/TNS.2010.2042615

Girard, S.; Ouerdane, Y.; Marcandella, C.; Boukenter, A.; Quenard, S. & Authier, N. (2011). Feasibility of radiation dosimetry with phosphorus-doped optical fibers in the ultraviolet and visible domain, *J. Non-Cryst. Solids*, Vol. 357, (2011), (1871–1874), doi:10.1016/j.jnoncrysol.2010.11.113

Goettmann, W.; Wulf, F.; Körfer, M. & Kuhnhenn, J. (2005). Beam loss position monitor using Cerenkov radiation in optical fibers, *Proceedings of the 7th European Workshop on Beam Diagnostics and Instrumentation for Particle Accelerators*, pp. 301-303, Lyon, June 2005

Goettmann, W.; Körfer, M. & Wulf, F. (2007). Beam profile measurement with optical fiber sensors at FLASH, *Proceedings of 8th European Workshop on Beam Diagnostics and Instrumentation for Particle Accelerator*, pp. 123-125, Venice, May 2007

Griscom, D.L.; Golant, K.M.; Tomashuk, A.L.; Pavlov, D.V. & Tarabrin, Yu.A. (1996). γ-radiation resistance of aluminum-coated all-silica optical fibers fabricated using different types of silica in the core, *Appl. Phys. Lett.*, Vol. 69, (1996), (322-324)

Gusarov[a], A.I.; Starodubov, D.S.; Berghmans, F.; Deparis, O.; Defosse, Y.; Fernandez Fernandez, A.; Decréton, M.; Mégret, P. & Blondel, M. (1999). Comparative study of MGy dose level g-radiation effect on FBGs written in different fibres, presented at *OFS-13*, Kyongju, April 1999

Gusarov[b], A. I.; Berghmans, F.; Deparis, O.; Fernandez Fernandez, A.; Defosse, Y.; Mégret, P.; Décreton, M. & Blondel, M. (1999), High total dose radiation effects on temperature sensing fiber bragg gratings, *IEEE Photon. Tech. L.*, Vol. 11, No. 9, (September 1999), (1159-1161)

Gusarov, A.; Fernandez Fernandez, A.; Vasiliev, S.; Medvedkov, O.; Blondel, M. & Berghmans, F. (2002). Effect of gamma-neutron nuclear reactor radiation on the properties of Bragg gratings written in photosensitive Ge-doped optical fiber, *Nucl. Instr. Met. Phys. Res.*, Vol. B187, (2002), (79-86)

Gusarov[a], A.; Vasiliev, S.; Medvedkov, O.; Mckenzie, I. & Berghmans, F. (2008). Stabilization of fiber Bragg gratings against gamma radiation, *IEEE T. Nucl. Sci.*, Vol. 55, No. 4, (2008), 2205 – 2212, doi: 10.1109/TNS.2008.2001038

Gusarov[b], A.; Chojetzki, C.; Mckenzie, I.; Thienpont, H. & Berghmans, F. (2008). Effect of the fiber coating on the radiation sensitivity of type I FBGs, *IEEE Photon. Tech. L.*, Vol. 20, No. 21, (November, 2008), (1802-1804), doi: 10.1109/LPT.2008.2004699

Gusarov, A.; Brichard, B. & Nikogosyan, D.N. (2010). Gamma-radiation effects on Bragg gratings written by femtosecond UV laser in Ge-doped fibers, *IEEE T. Nucl. Sci.*, Vol. 57, No. 4, (2010), (2024 – 2028), doi: 10.1109/TNS.2009.2039494

Hanafusa, H.; Hibino, Y. & Yamamoto, F. (1986). Drawing condition dependence of radiation-induced loss in optical fibres, *Electron. Lett.*, Vol. 22, No. 2, (1986), (106-108).

Hashim, S.; Ramli, A.T.; Bradley, D.A. & Wagiran,, H. (2008). The thermoluminescence response of Ge-doped optical fibers to X-ray photon irradiation, *J. Fiz. UTM.*, Vol. 3, (2008), (31-37)

Henschel, H.; Köhn, O. & Schmidt, H.U. (1992). Optical fibres as radiation dosimeters, *Nucl. Instr. Met. Phys. Res.*, Vol. B69, (1992), (307-314)

Henschel, H.; Körfer, M.; Wittenburg, K. & Wulf, F. (2000). Fiber optic radiation sensing systems for TESLA, *TESLA Report No. 2000-26*, September 2000

Henschel, H.; Körfer, M.; Kuhnhenn, J.; Weinand, U. & Wulf, F. (2004). Fibre optic radiation sensor systems for particle accelerators, *Nucl. Instr. Met. Phys. Res.*, Vol. A526, (2004), (537-550), doi:10.1016/j.nima.2004.02.030

Huston, A.L.; Justus, B.L. & Johnson, T.L. (1996). Fiber-optic-coupled, laser heated thermoluminescence dosimeter for remote radiation sensing, *Appl. Phys. Lett.*, Vol. 68, No. 24, (June 1996), (3377-3379)

Ichikawa, T., Mechanism of radiation-induced degradation of poly(methyl methacrylate) -- temperature effect, *Nucl. Instr. Met. Phys. Res. Section B: Beam Interactions with Materials and Atoms*, Vol. 105, No. 1-4, (1995), (150-153), doi:10.1016/0168-583X(95)00534-X

Intermite, A.; Putignano, M. & Welsch, C. P. (2009). Feasibility study of an optical fiber sensor for beam loss detection based on a SPAD array, *Proceedings of 9th European Workshop on Beam Diagnostics and Instrumentation for Particle Accelerators*, pp. 228-230, Basel, May 2009

Jang, K.W.; Cho, D.H.; Shin, S.H.; Yoo, W.J.; Seo, J.K.; Lee, B.; Kim, S.; Moon, J.H.; Cho, Y.-H. & Park, B.G. (2009). Characterization of a Scintillating Fiber-optic Dosimeter for Photon Beam Therapy, *Opt. Rev.*, Vol. 16, No. 3, (2009), (383–386)

Jang[a], K.W.; Cho, D.H.; Yoo, W.J.; Seo, J.K.; Heo, J.Y.; Lee, B.; Cho, Y.-H.; Park, B.G.; Moon, J.H. & Kim, S. (2010). Measurement of Cerenkov light in a fiber-optic radiation sensor by using high-energy photon and electron beams, *J. Korean Phys. Soc.*, Vol. 56, No. 3, (March 2010), (765-769), doi: 10.3938/jkps.56.765

Jang[b], K.W.; Yoo, W.J.; Park, J. & Lee, B. (2010). Development of scintillation-fiber sensors for measurements of thermal neutrons in mixed Neutron-gamma fields, *J. Korean Phys. Soc.*, Vol. 56, No. 6, (June 2010), (1777-1780), DOI: 10.3938/jkps.56.1777

Jang[c], K.W.; Cho, D.H.; Yoo, W.J.; Seo, J.K.; Heo, J.Y.; Park, J.-Y. & Lee, B. (2010). Fiber-optic radiation sensor for detection of tritium, *Nucl. Instr. Methods Phys. Res.*, doi:10.1016/j.nima.2010.09.060

Jang[a], K.W.; Yoo, W.J.; Seo, J.K.; Heo, J.Y.; Moon, J.; Park, J.-Y.; Kim, S.; Park, B.G. & Lee, B. (2011). Measurements and Removal of Cerenkov light generated in scintillating fiber-optic sensor induced by high-energy electron beams using a spectrometer, *Opt. Rev.*, Vol. 18, No. 1, (2011), (176–179)

Jang[b], K.W.; Lee, B. & Moon, J.H. (2011). Development and characterization of the integrated fiber-optic radiation sensor for the simultaneous detection of neutrons and gamma rays, *Appl. Radiat. Isotopes*, Vol. 69, (2011) (711–715), doi:10.1016/j.apradiso.2011.01.009

Jones, A.K. & Hintenlang, D. (2008). Potential clinical utility of a fiber optic-coupled dosemeter for dose measurements in diagnostic radiology, *Radiat. Prot. Dosim.*, Vol. 132, No. 1, (2008), (80–87), doi:10.1093/rpd/ncn252

Jones, S.C.; Hegland, J.E.; Hoffman, J.M. & Braunlich, P.F. (1990). A fibre-optic TLD microprobe for remote in-vivo radiotherapy dosimetry, *Radiat. Prot. Dosim.*, Vol. 34, No. 1/4, (1990), (279-282)

Jones, S.C.; Sweet, J.A.; Braunlich, P.; Hoffman, J.M. & Hegland, J.E. (1993). A remote fibre optic laser TLD system, *Radiat. Prot. Dosim.*, Vol. 47, No. 1/4, (1993), (525-528)

Justus, B.L.; Falkenstein, P.; Huston, A.L.; Plazas, M.C.; Ning, H. & Miller, R.W. (2006). Elimination of Cerenkov interference in a fibre-optic-coupled radiation dosimeter, *Radiat. Prot. Dosim.*, Vol. 120, No. 1–4, (2006), (20–23), doi:10.1093/rpd/nci525

Klein, D.M. & McKeever, S.W.S. (2008). Optically stimulated luminescence from KBr:Eu as a near-real-time dosimetry system, *Radiat. Meas.*, Vol. 43, (2008), (883 – 887), doi:10.1016/j.radmeas.2008.01.015

Krebber, K., Henschel, H. & Weinand, U. (2006). Fibre Bragg gratings as high dose radiation sensors?, *Meas. Sci. Technol.*, Vol. 17, (2006), (1095–1102), doi:10.1088/0957-0233/17/5/S26

Kuhnhenn, J.; Henschel, H.; Köhn, O. & Weinand, U. (2004). Thermal annealing of radiation dosimetry fibres, *Proceedings RADECS 2004*, pp. 39-42, Madrid, September 2004

Kuhnhenn, J. (2005). Radiation tolerant fibres for LHC controls and communications, *5th LHC Radiation Workshop*, CERN, November 2005

Kuyt, G.; Regnier, E. & Gilberti, R. (2006). Optical fiber behavior in radioactive environments, presented at *IEEE ICC meeting*, St-Peterburg, 2006

Lee[a], Bongsoo; Jang, Kyoung Won; Cho, Dong Hyun; Yoo, Wook Jae; Kim, Hyung Shik; Chung, Soon-Cheol & Yi, Jeong Han (2007). Development of one-dimensional fiber-optic radiation sensor for measuring dose distributions of high energy photon beams, *Opt. Rev.*, Vol. 14, No. 5 (2007) (351–354)

Lee[b], Bongsoo; Jang, Kyoung Won; Cho, Kyoung Won; Yoo, Wook Jae; Tack, Gye-Rae; Chung, Soon-Cheol; Kim, Sin & Cho, Hyosung. (2007). Measurements and elimination of Cherenkov light in fiber-optic scintillating detector for electron beam therapy dosimetry, *Nucl. Instr. Methods Phys. Res.* Vol. A579, (2007), (344-348), doi:10.1016/j.nima.2007.04.074

Liu, Y.-P.; Chen, Z.-Y.; Ba, W.-Z.; Fan, Y-W.; Du, Y.-Z.; Pan, S.-L. & Guo, Q. (2008). A study on the real-time radiation dosimetry measurement system based on optically stimulated luminescence, *Chinese Phys. C (HEP & NP)*, Vol. 32, No. 5, (May 2008), (381-384)

Lopez-Higuera, J. M. (1998). *Optical Sensors*, Universidad de Cantabria

Lu, P.; Bao, X.; Kulkarni, N. & Brown, K. (1999). Gamma ray radiation induced visible light absorption in P-doped silica fibers at low dose levels, *Radiat. Meas.*, Vol. 30, (1999), (725-733)

Mady, F.; Benabdesselam, M., Mebrouk, Y. & Dussardier, B. (2010). Radiation effects in ytterbium-doped silica optical fibers: traps and color centers related to the radiation-induced optical losses, *RADECS 2010 Proceedings*, Paper LN2, http://hal.archives-ouvertes.fr/hal-00559422/en/

Magne, S.; De Carlan,L.-L.; Sorel, S.; Bordy, J.-M.; Isambert, A. & Bridier, A. (2008). MAESTRO Project: evelopment of a multi-Channel OSL dosimetricsystemfor clinical use, *Workshop on dosimetric issues in the medical use of ionizing radiation EURADOS AM 2008*, Paris, January 2008

Maier, R. R.J.; MacPherson, W.N.; Barton, J.S.; Jones, J.D.C.; McCulloch, S., Fernandez-Fernandez, A.; Zhang, L. & Chen, X. (2005). Fibre Bragg gratings of type I in SMF-28 and B/Ge fibre and type IIA B/Ge fibre under gamma radiation up to 0.54 MGy,

Proceedeidngs of the 17th International Conference on Optical Fibre Sensors, Voet, M., Willsch, R., Ecke, W., Jones, J. & Culshaw, B. , SPIE Vol. 5855, pp. 511-514, 2005, doi: 10.1117/12.624037

Mao, R.; Zhang, L. & Zhu, R-Y. (2009). Gamma ray induced radiation damage in PWO and LSO/LYSO crystals, *Proceedings of the 2009 IEEE Nuclear Science Symposium Conference Record*, (2009), (2045-2049), http://authors.library.caltech.edu/19580/1/Mao2009p111452008_Ieee_Nuclear_Science_Symposium_And_Medical_Imaging_Conference_(2008_NssMic)_Vols_1-9.pdf

McCarthy, D.; O'Keeffe, S. & Lewis, E. (2010). Optical fibre radiation dosimetry for low dose applications, presented at *IEEE Sensors Conference 2010*, October 2010, Hawaii

McCarthy, D.; O'Keeffe, S.; Sporea, D.; Sporea, A.; Tiseanu, I. & Lewis, E. (2011). Optical Fibre X-Ray Radiation Dosimeter Sensor for Low Dose Applications, presented at *IEEE Sensors Conference 2011*, October 2011, Limerick

McLaughlin, W.L.; Miller, A .; Boyd, A.W.; McDonald, J. C. & Chadwick, K.H. (1989). *Dosimetry for Radiation Processing*, Taylor & Francis, 1989, ISBN-13: 9780850667400

Merlo, J.P & Cankoçak, K. (2006). Radiation-hardness studies of high OH−content quartz fibers irradiated with 24 GeV protons, *CMS Conference Report*, January 2006, http://cdsweb.cern.ch/record/926539/files/CR2006_005.pdf

Miniscalco, W.J.; Wei, T. & Onorato, P.K. (1986). Radiation hardened silica-based optical fibers, *RADC-TR-88-279 - Final Technical Report December 1986*, Rome Air Development Center, Air Force Systems Command, GTE Laboratories

Moloney, W.E. (2008). *A fiber-optic coupled point dosimetry system for the characterization of multi-detector computer tomography*, Master of Science Thesis, University of Florida, 2008

Mommaert, C. (1992). Optoelectronic readout for scintillating fiber trackers, *Nucl. Instr. Met. Phys. Res.*, Vol. A323, (1992), (477-484)

Naka, R.; Watanabe, K.; Kawarabayashi, J.; Uritani, A.; Iguchi, T.; Hayashi, N.; Kojima, N.; Yoshida, T.; Kaneko, J.; Takeuchi, H. & Kakuta, T. (2001). Radiation distribution sensing with normal optical fiber, *IEEE T. Nucl. Scie.*, Vol. 48, No. 6, (December 2001), (2348-2351)

Nakajima, D.; Őzel-Tashenov, B.; Bianchin, S.; Borodina, O.; Bozkurt, V.; Göküzüm, B.; Kavatsyuk, M.; Minami, S.; Rappold, C.; Saito, T.R.; Achenbach, P.; Ajimura, S.; Ayerbe, C.; Fukuda, T.; Hayashi, Y.; Hiraiwa, T.; Hoffmann, J.; Koch, K.; Kurz, N.; Lepyoshkina, O.; Maas, F.; Mizoi, Y.; Mochizuki, T.; Moritsu, M.; Nagae, T.; Nungesser, L.; Okamura, A.; Ott, W.; Pochodzalla, J.; Sakaguchi, A.; Sako, M.; Schmidt, C.J.; Sugimura, H.; Tanida, K.; Träger, M.; Trautmann, W. & Voltz, S. (2009). Scintillating fiber detectors for the HypHI project at GSI, *Nucl. Instr. Met. Phys. Res.*, Vol. A608, (2009), (287–290), doi:10.1016/j.nima.2009.06.068

Nishiura, R. & Izumi, N. (2001). Radiation sensing system using an optical fiber, *Mitsubishi Electric ADVANCE*, September 2001, pp. 25-28

Okamoto, K.; Toh, K.; Nagata, S.; Tsuchiya, B.; Suzuki, T.; Shamoto, N. & Shikama, T. (2004). Temperature dependence of radiation induced optical transmission loss in fused silica core optical fiber, *J. Nucl. Mater.*, Vol. 329–333, (2004), (1503–1506), doi:10.1016/j.jnucmat.2004.04.243

O'Keeffe, S.; Fernandez Fernandez, A.; Fitzpatrick, C.; Brichard, B. & Lewis, E. (2007). Real-time gamma dosimetry using PMMA optical fibres for applications in the sterilization industry, *Meas. Sci. Technol.*, Vol. 18, No. 10, (2007), (3171-3176)

O'Keeffe, S.; Fitzpatrick, C.; Lewis, E. & Al-Shamma'a, A.I. (2008). A review of optical fibre radiation dosimeters, *Sensor Rev.*, Vol. 28, No. 2, (2008), (136–142), doi 10.1108/02602280810856705

O'Keeffe, S. & Lewis, E. (2009). Polymer optical fibre for in situ monitoring of gamma radiation processes, *Intl. J. Smart Sensing and Intelligent Systems*, VOL. 2, No. 3, (September 2009), (490-502)

O'Keeffe[a], S.; Lewis, E.; Santhanam, A. & Rolland, J.P. (2009), Variable Sensitivity Online Optical Fibre Radiation Dosimeter, presented at *The Eighth IEEE Conference on Sensors (IEEE Sensors Conference)*, October 2009, Christchurch

O'Keeffe[b], S.; Lewis E.; Santhanam, A.; Winningham, A. & Rolland, J.P. (2009). Low dose plastic optical fibre radiation dosimeter for clinical dosimetry applications, presented at *The Eighth IEEE Conference on Sensors (IEEE Sensors Conference)*, October 2009, Christchurch

Ong, C.L.; Kandaiya, S.; Kho, H.T. & Chong, M.T. (2009). Segments of a commercial Ge-doped optical fiber as a thermoluminescent dosimeter in radiotherapy, *Radiat. Meas.*, Vol. 44, (2009), (158–162), doi:10.1016/j.radmeas.2009.01.011

Origlio, G. (2009). *Properties and radiation response of optical fibers: role of dopants*, PhD Thesis, http://portale.unipa.it/export/sites/www/dipartimenti/fisica/home/attachments/tesi_dottXXI/Origlio_PhD.pdf

Paul, M.C.; Bohra, D.; Dhar, A.; Sen, R.; Bhatnagar, P.K. & Dasgupta, K. (2009). Radiation response behavior of high phosphorous doped step-index multimode optical fibers under low dose gamma irradiation, *J. Non-Cryst. Solids*, Vol. 355, (2009), (1496-1507), doi:10.1016/j.jnoncrysol.2009.05.017

Pittet, P.; Lua, Guo-Neng; Galvana, J.-M.; Loisya, J.-Y.; Ismaild, A.; Giraudd, J.-Y. & Balossod, J. (2009). Implantable real-time dosimetric probe using GaN as scintillation material, *Sensors Actuators A: Physical*, Vol. 151, (2009), (29–34), doi:10.1016/j.sna.2009.02.018

Plazas, M. C.; Justus, B. L.; Falkenstein, P.; Huston, A. L.; Ning, H. & Miller, R. (2005). Optical fiber detectors as in-vivo dosimetry method of quality assurance in radiation therapy, *Revista Colombiana de Física*, Vol. 37, No. 1, (2005), (307-313)

Primak, W.; Edwards, E.; Keiffer, D. & Szymansk, H. (1964). Ionization expansion of compacted silica and the theory of radiation-induced dilatations in vitreous silica, *Phys. Rev.*, Vol. 133, No. 2A, (January 1964), (A531-A535)

Pruett, B. L.; Peterson, R. T.; Smith, D. E.; Looney, L. D. & Shelton, Jr., R. N. (1984). Gamma-ray to Cerenkov-light conversion efficiency for pure-silica-core optical fibers, *SPIE Proceedings*, Vol. 506, pp. 10-16, San Diego, August 1984

Radiation effects, The 10 th Europhysical Conference on Defects in Insulation Materials, *Phys. Status Solidi*, Vol. 4, No. 3 , (2007), (1060-1175)

Regnier, E.; Flammer, I.; Girard, S.; Gooijer, F.; Achten, F. & Kuyt, G. (2007). Low-dose radiation-induced attenuation at Infrared wavelengths for P-doped, Ge-doped and pure silica-core optical fibres, *IEEE T. Nucl. Sci.*, Vol. 54, No. 4, (August 2007), (1115-1119),   doi: 10.1109/TNS.2007.894180,

http://indico.cern.ch/getFile.py/access?contribId=1&resId=1&materialId=0&conf
Id=37455

Rüdiger, F.; Körfer, M.; Göttmann, W.; Schmidt, G. & Wille, K. (2008). Beam loss position
monitoring with optical fibres at DELTA, *Proceedings of EPAC 2008*, pp. 1032-1034,
Genoa, June 2008

Saint-Gobain Crystals and Detectors (2005). *Scintillating Optical Fibers*,
http://www.detectors.saint-
gobain.com/uploadedFiles/SGdetectors/Documents/Brochures/Scintillating-
Optical-Fibers-Brochure.pdf

Santiago, M.; Prokic, M.; Molina, P.; Marcazzó, J. & Caselli, E. (2009). A tissue-equivalent
radioluminescent fiberoptic probe for in-vivo dosimetry based on Mn-doped
lithium tetraborate, *WC 2009, IFMBE Proceedings 25/ III*, Dössel, O. & Schlegel, W C.
(Eds.), pp. 367–370, 2009

Seo, M.W; Kim, J.K. & Park, J.W. (2009). Gamma-ray induced radiation M. W damage in a
small size inorganic scintillator, *The Tenth International Conference on Inorganic
Scintillators and their Applications SCINT 2009*, Jeju, June 2009

Seo, M.W; Kim, J.K. & Park, J.W. (2011). Test of a fiber optic scintillation dosimeter with
BGO tip in a ^{60}Co irradiation chamber, *Progress in Nucl. Scie. and Techn.*, Vol. 1,
(2011), (186-189)

Shikama, T.; Toh, K.; Nagata, S.; Tsuchiya, B.; Yamauchi, M.; Nishitani, T.; Suzuki, T.;
Okamoto, K.; Kubo, N.; Ishihara, M. & Kakuta, T. (2004). Optical fast neutron and
gamma-ray detection by radioluminescence, *JAERI-Review*, (2004-017), (101-105)

Skuja, L; Tanimura, K & Itoh, N. (1996). Correlation between the radiation-induced intrinsic
4.8 eV optical absorption and 1.9 eV photoluminescence bands in glassy SiO_2, *J.
Appl. Phys.*, Vol. 80, No. 6, (September 1996), (3518 – 3525), doi: 10.1063/1.363224

Sporea, D. & Sporea, R. (2005). Setup for the in situ monitoring of the irradiation-induced
effects in optical fibers in the ultraviolet-visible optical range, *Rev. Sci. Instrum.*, Vol.
76, No. 11, (2005), doi:10.1063/1.2130932

Sporea, D. & Sporea, A. (2007). Radiation effects in sapphire optical fibers, Phys. Status
Solidi (c), Vol. 4, No. 3, (2007), (1356–1359), doi: 10.1002/pssc.200673709

Sporea[a], D., Agnello, S. & Gelardi, F.M. (2010). Irradiation Effects in Optical Fibers, In:
Frontiers in Guided Wave Optics and Optoelectronics, Pal, B., pp. 49-66, Intech, ISBN
978-953-7619-82-4, Vienna, Austria

Sporea[b], D.; Sporea, A.; Oproiu, C.; Vata, I.; Negut, D.; Secu, M & Grecu, M. (2010). Annual
Report, Contract 12084/2008

Suchowerska, N.; Lambert, J.; Nakano, T.; Law, S.; Elsey, J. & McKenzie, D.R. (2007). A fibre
optic dosimeter customised for brachytherapy, *Radiat. Meas.*, Vol. 42, (2007), (929–
932), doi:10.1016/j.radmeas.2007.02.042

Takada, E.; Kimura, A.; Hosono, Y.; Takahashi, H. & Nakazawa, M. (1999). Radiation
distribution sensor with optical fibers for high radiation fields, *J. Nucl. Sci. Techn.*,
Vol. 36, No. 8, (August 1999), (641-645)

Tanyi, J. A.; Nitzling, K.D.; Lodwick, C.J.; Huston, A.L. & Justus, B.L. (2011).
Characterization of a gated fiber-optic-coupled detector for application in clinical
electron beam dosimetry, *Med. Phys.*, Vol. 38, No. 2, (February 2011), (961-967), doi:
10.1118/1.3539737

Thériault, S. (2006). Radiation effects on COTS laser-optimized multimode fibers exposed to an intense gamma radiation field, presented at *Photonics North*, Quebec City, June 2006

Toh, K.; Shikama, T.; Kakuta, T.; Nagata, S.; Tsuchiya, B.; Mori, C.; Hori, J. & Nishitani, T. (2002). Studies of radioluminescence in fused silica core optical fibers, *Proc. SPIE*, Vol. 4786, (2002), doi:10.1117/12.451745

Toh, K.; Shikama, T.; Nagata, S.; Tsuchiya, B.; Suzuki, T.; Okamoto, K.; Shamoto, N.; Yamauchi, M. & Nishitani, T. (2004). Optical characteristics of aluminum coated fused silica core fibers under 14 MeV fusion neutron irradiation, *J. Nucl. Mater.*, Vol. 329–333, (2004), (1495–1498), doi:10.1016/j.jnucmat.2004.04.245

Treadaway, M.J.; Passenheim, B.C. & Kitterer, B.D. (1975). Luminescence and absorption of electron-irradiated common optical glasses, sapphire, and quartz, IEEE *T. Nucl. Sci.*, Vol.22, No. 6, (1975), (2253-2258), doi: 10.1109/TNS.1975.4328115

Tremblay, N.M.; Hubert-Tremblay, V.; Rachel Bujold, Rachel; Beaulieu, L. & Lepage, M. (2010). Improvement in the accuracy of polymer gel dosimeters using scintillating fibers, *J. Phys. Conf. Ser.*, Vol. 250, (2010), (1-4), doi:10.1088/1742-6596/250/1/012076

Tsuchiya, B.; Kondo, S.; Tsurui, T.; Toh, K.; Nagata, S. & Shikama, T. (2011). Correlation between radiation-induced defects, and optical properties of pure fused silica-core optical fiber, under gamma-ray irradiation in air at 1273 K, *J. Nucl. Mater.*, (2011), in press, doi:10.1016/j.jnucmat.2010.12.164

Vedda, A.; Chiodini, N.; Di Martino, D.; Fasoli, M.; Keffer, S.; Lauria, A.; Martini, M.; Moretti, F.; Spinolo, G.; Nikl, M.; Solovieva, N. & Brambilla, G. (2004). Ce3+-doped fibers for remote radiation dosimetry, *Appl. Phys. Lett.*, Vol. 85, No. 26, (December 2004), (6356–6358), doi: 10.1063/1.1840127

Weeks, R. A. & Sonder, E. (1963). The relation between the magnetic susceptibility, electron spin resonance, and the optical absorption of the E1'center in fused silica, In: *Paramagnetic Resonance II*, W. Low, pp. 869-879, Academic Press, LCCN 63-21409, New York

Weinert, A. (1999) *Plastic Optical Fibers*, Publicis MCD Verlag

Wijnands, T.; De Jonge, L.K.; Kuhnhenn, J.; Hoeffgen, S. K. & Weinand, U. (2007). Radiation tolerant optical fibres for LHC beam instrumentation, presented at the *6th LHC Radiation Workshop*, CERN, September 2007

Wittenburg, K. (2008). Beam loss monitors, http://cdsweb.cern.ch/record/1213279/files/p249.pdf

Wrobel, F. (2005). Fundamentals on Radiation-Matter Interaction, *New Challenges for Radiation Tolerance Assessment, RADECS 2005 Short Course Notebook*, Cap d'Agde, September 2005

Wu, F.; Ikram, K. & Albin, S. (2003). Photonic Crystal Fiber for Radiation Sensors, presented at *Workshop on Innovative Fiber Sensors*, Newport News, May 2003

Wulf, F. & Körfer, M. (2009). Beam loss and beam monitoring with optical fibers, *Proceedings of 9th European Workshop on Beam Diagnostics and Instrumentation for Particle Accelerators*, pp. 411-417, Basel, May 2009

Yaakob, N.H.; Wagiran, H.; Hossain, I.; Ramli, A.T.; Bradley, D.A.; Hashim, S.; Ali, H. (2011). Electron irradiation response on Ge and Al-doped SiO₂ optical fibres, *Nucl. Instr. Met. Phys. Res.*, Vol. A637, (2011), (185–189), doi:10.1016/j.nima.2011.02.041

Yoshida, H. & Ichikawa, T. (1995). Temperature effect on the radiation-degradation of poly(methyl methacrylate), *Radiat. Phys. Chem.*, Vol. 46, No. 4-6, Part 1, (1995), (921–924), doi:10.1016/0969-806X(95)00293-7

Yoshida, Y.; Muto, S. & Tanabe, T. (2007). Measurement of soft X-ray excited optical luminescence of a silica glass, In: *CP882, X-ray Absorption Fine Structure – XAFS73*, Hedman, B. & Pianetta, P., 2007, American Institute of Physics

Yukihara, E.G.; Sawakuchi, G.O.; Guduru, S.; McKeever, S.W.S.; Gaza, R.; Benton, E.R.; Yasuda, N.; Uchihori, Y. & Kitamura, H. (2006). Application of the optically stimulated luminescence (OSL) technique in space dosimetry, *Radiat. Meas.*, Vol. 41, (2006), (1126 – 1135), doi:10.1016/j.radmeas.2006.05.027

10

Nanoparticles On A String – Fiber Probes as "Invisible" Positioners for Nanostructures

Phillip Olk
NTNU Trondheim
Norway

1. Introduction

Optical fibers have a long and successful history in telecommunications (Howes & Morgan, 1980) and had an enormous global impact on technology and cultural interaction, far beyond just taking telegraphy further (Stephenson, 1996). Aside from transporting data from A to B, optical fibers allowed the exploration of a whole field in photonics research, from fiber-based sensors (Udd, 1990) via various photonic crystal fibers (Zolla et al., 2005) up to quantum optical experiments (Philbin et al., 2008) - current research is presented in this book.

In this chapter, optical fibers are used just as a pointed probe which is basically transparent for light. So most of the originally engineered waveguide properties are ignored, such as low dispersion and absorption for long range applications. Contrary, using optical fibers as scanning probe tips exploits optical and mechanical effects on the nano-scale, i.e., of very close proximity, even less than 5 nm. This is a well-established "abuse" of telecommunication fibers in the Scanning Nearfield Optical Microscope[1], an opto-mechanical tool that is of high benefit in the field of nano-optics (Dunn, 1999). These scanning probe optical microscopes exploit one property of optical fibers that is often overlooked, as it's taken as granted: optical fibers are *transparent*.

 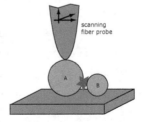

Fig. 1. *Left*. Numerical calculations of the intensity in the vicinity of two metal spheres A and B (Au, 80 nm and 30 nm in diameter, respectively), exposed to a plane wave (λ =532 nm in vacuum). The hot spot has a 20x relative intensity w/respect to the incident wave. *Right*. Suggested experimental implementation: particle A is glued to a 3D-positionable (potentially rotatable) fiber probe, whereas particle B is resting on the substrate.

---

[1] Abbreviated SNOM, or NSOM, depending on Your current position on the globe.

In the field of nano-optics, the effects of near-fields are applied in order to manipulate light on a scale distinctively smaller than the wavelength of that light. One handy effect is the so-called near-field enhancement at metal structures: a tiny metal rod , i.e. less than $\lambda/4$ in length and only a few dozens of nm in diameter, or a metal sphere can serve as dipole antenna when exposed to a propagating electromagnetic field. The electrons just follow the external electrical field which is oscillating with the frequency of light, and if metal geometry, light frequency and electron mobility/velocity are well trimmed, the particle may exhibit a plasma resonance. If one examines the electric field in the very vicinity of the dipole one will encounter local intensities which are higher than the incident field. In other words: a nanometer-scaled lump of metal may serve as an *optical antenna* and provides enhanced local intensities. As the volume of enhanced intensity is rather small, such a construction can be used as a highly localized light source. E.g., this is applied in tip-enhanced Raman spectroscopy (TERS), where "intensity concentration" is crucial in order to increase the Raman response, which tends to have a very small cross section, but scales with I^2.

Fig.1 depicts a basic application idea of such a nanoantenna: as calculations hint, Fig.1 *left*, a local field enhancement may exist between two nanoparticles, which in turn might be used for experiments such as fluorescence excitation, Raman scattering and similar. In Fig.1 *right*, a sketch is given for a real experimental implementation of such a bi-particle nano-antenna. One particle is attached to a scanning probe, while the second is resting on a substrate. As the actual local field enhancement is highly dependent on the inter-particle distance and the relative orientation to the polarisation, the scanning fiber tip is used as a manipulator in order to optimise the system. Note that no optical properties of the tip are used.

The following sections will provide a detailed insight into the methods of mounting a such single metal nanoparticle to a scanning probe. The reader might keep in mind that the concepts here are not restricted to optical antennas on fiber tips, but can be transferred to different materials for both probe (glass, Si, metals...) and nanoparticle (metal, luminescent nanodiamonds, micelles, cells, molecules etc.), and thus into other wavelength regimes (Wenzel et al., 2008). For the sake of simplicity, only metal nanoparticles are discussed here in detail.

While the idea of, e.g., particle-enhanced Raman spectroscopy is older than the expression "optical antenna" (Wessel, 1985), it turned out to be rather difficult to determine an appropriate nano-particle and subsequently attach it to a scanning probe fulfilling these conditions:

- Exactly the pre-selected particle shall be attached, not just some random item.

- Only this single one shall be attached, not "two, maybe three".

- The particle (especially its optical properties) shouldn't be altered by the pick up process.

- The position and orientation of the attached particle on the probe tip must be known exactly afterwards.

A scientist who is only limited by "technical feasibility" might think of manufacturing techniques such as electron beam lithography: evaporate metal on a glass needle, cut away the surplus metal, done. This certainly is a valid approach, but not everybody has, wants, likes to use, or can afford such an e-beam tool. The technique, as largely introduced by (Kalkbrenner et al., 2001) and continued here, is relatively cheap: if one can afford a SNOM and a usual optical microscope, particle-decorated tips are well within reach.

Most certainly, alternative experiments can generate very similar information about the system of interest. For example, distance-dependent spectroscopy of particle pairs can be realized by producing vast arrays of particle pairs by some lithography technique (Rechberger et al., 2003), or by pushing single particles into positions by means of a scanning probe (Bek et al., 2008). These techniques are well suitable for two-dimensional positioning, but an experimentalist may encounter a situation where she or he wants to have full control of both position and orientation of a three-dimensional particle in a three-dimensional space – this can be achieved by moving and rotating the particle-decorated scanning probe (Kalkbrenner et al., 2004). As the fiber, especially if immersed in index matching fluid, is hardly visible, the comparison to marionette string puppets is self-evident.

2. Manufacturing particle-decorated fiber tips

This section describes a workflow for attaching a particle to a scanning probe tip for further experiments. Starting from common basics, a few details were optimised, but the procedure may need adaptation for given instruments and circumstances.

2.1 The basic probe design

As a start, an "empty" scanning fiber probe is needed. Depending on time, budget and capabilities, one may either buy a ready-made one or build it from scratch using a raw optical fiber, a quartz tuning fork, and glue. This fiber tip is part of a scanning probe system. Depending on the actual setup and experiment, various mechanisms and realisations are imaginable, Fig. 2, and most of them were successfully applied for particle decoration experiments (Kalkbrenner et al., 2001; Olk et al., 2007; Uhlig et al., 2007). For mid-IR and THz applications, even silicon is more or less transparent, so the SNOM tip may be replaced by a slightly modified AFM cantilever (Wenzel et al., 2008).

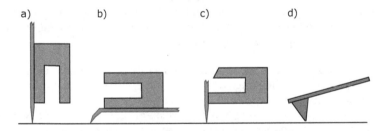

Fig. 2. Variations of scanning probe realizations. a)-c) are based on fiber tips, glued to common quartz tuning forks. d) is a usual AFM tip (Si), which may be sufficiently transparent for IR and THz experiments.

Independent on how the scanning probe is obtained, it is actually desired to have a probe tip *not* as pointed as possible: the tip should not be smaller than the particle that is to be picked. This "trick" allows much bigger tolerances when aligning and positioning the scanning probe with respect to the nanoparticle, Fig.3. In addition, this facilitates the manufacturing process of the pointed fiber itself, as it is not mandatory to use HF acid in order to etch the glass: a heat-and-pull technique, as used in micropipette pullers, can be sufficient.

2.2 Particle decoration

This section provides an all-optical method to pre-select, attach, and check the successful attachment of a desired particle. Depending on the actual experiment, some steps can be

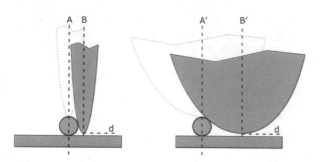

Fig. 3. Tolerances when picking a given particle with a pointed (left) or blunt tip (right). The ideal tip position is hinted by the dotted outlines, and is symmetric to the line A or A', respectively. For a given acceptable minimum vertical tip-sample-distance d, the lateral tolerance is small for pointed tips, $\overline{AB}$, and significantly bigger for blunt tips, $\overline{A'B'}$.

omitted (e.g., the orientation check for spherical particles) or might need to be added (e.g. time-correlated photon counting to check the number of emitters in a nanocrystal).

2.2.1 Particle preparation
Nanoparticles are usually provided as powder, or as suspension in a liquid. For the latter case, a single droplet is applied onto a glass cover slide, as common in microscopy. Depending on the particle density in the suspension liquid, this droplet may just dry (for very low densities), or be sent across the substrate for tens of seconds by a hand-operated rubber blower. As a result, single nanoparticle "candidates" should be distributed over the slide with a next-neighbour-distance big enough so a) single particles can be analysed optically and b) the more or less blunt fiber tip will pick up only one particle.

2.2.2 Optical microscopy of single nanoparticles – the role of homogeneous immersion
For standard microscopy, the Rayleigh criterion is still a valid limit for the resolution. Nonetheless, standard optics are capable to "see" nanoscopic particles:
According to Mie's theory of scattering and absorption of light (plane wave, wavelength λ) at a small (radius $R < \lambda$) conductive sphere, the scattering cross section is proportional to $C_{scat} \propto R^6$, or, as volume $V = 4/3\pi R^3$:

$$C_{scat} \propto V^2. \tag{1}$$

So even a very small particle is visible in the eyepiece of a microscope. It can't be *resolved* if close to an other one, but it will produce an Airy disk according to the point spread function of the microscope.[2]
For absorption, Mie provides a similar expression C_{abs}, but this is just proportional to the particle volume:

$$C_{abs} \propto V. \tag{2}$$

In a sketch of these two expressions, Fig.4 *left*, one realises that there exists a crossover size $V_\times$ for the particle, when the lucid scatterer ($C_{scat} > C_{abs}$ for $V > V_\times$) turns into an absorber ($V < V_\times$). For a real microscope situation as in Fig.5, this means that the observer faces a

---

[2] One can still tell single particles from doublets by both intensity and scattering spectrum, as will be explained later.

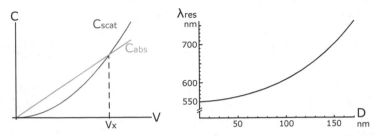

Fig. 4. Optical properties of conductive spheres. *Left.* Qualitative sketch of Mie's scattering cross section $C_{scat} \propto V^2$ and absorption cross section $C_{abs} \propto V$. They cross over at $V_\times$. *Right.* Spectral position λ_{res} of the plasma resonance of a particle of diameter D. The numerical values are valid for Au spheres, embedded in an effective medium of refraction $n = 1.52$, and a diameter range from 20 to 160 nm. λ_{res} is a vacuum wavelength, as measured by a (non-immersed) spectrometer.

contrast problem if the particle size is too small: big particles look bright, but in a transient size region, the residual reflection from the cover slide is about as bright as the scattering from the particle – "optical cloaking" without any cape involved. If the particle is even smaller, and the residual reflection is not too intense, the particles may be visible again, but this time, they are dark-on-bright ("bright" being the residual reflection), Fig.5b).

This situation is tightened for "modern standard" microscopy: immersion lenses easily provide numerical apertures NA > 1. This means that larger portions of the illumination light are reflected totally, as the maximum angle of incidence, φ_{NA} is easily bigger than the critical angel of incidence φ_{crit}, Fig.5b). As a result, the background intensity I_{bg} is elevated, so even a lucid scatterer is "swallowed" in contrast to the intense reflection. The solution to this is the homogeneous immersion, Fig.5c): by suppressing any reflection, only the scattering is observed. The idea is rather old (Abbe, 1873), but this is an example of lack of interdisciplinary exchange: scanning probe people prefer a dry environment for their probes, as this facilitates everything (Q factors, controlling etc.). On the other hand, life scientists, who deal with large NA objective lenses since many decades, never were aware of this problem, as they generally *always* use immersion fluids (mostly water).

Depending on the actual nanoparticle, other techniques might be applied in order to locate and identify single particles: absorption-mediated interference microscopy successfully imaged particles smaller than 2 nm (Boyer et al., 2002), and luminescent particles, such as fluorescent PS spheres or N-V centres in nanodiamonds, provide straight forward access to their location. Alternatively, if one succeeds in (self-)arranging nanoparticles in some fixed grid, only relative positioning of the sample is necessary.

A slightly different train of thoughts is valid for the scanning probe tip, which may be considered as a dielectric sphere. Here, absorption is not as important, but the index of refraction: the closer the indexes of the glass tip and the immersion medium are, the less light is "disturbed". Consequently, the tip can't be located by the bare eye. Indeed, the fiber core itself provides a variation of the index of refraction, an essential property for guiding optical modes, but this is too small to give a significant visual hint on the whereabouts of the fiber.

If there was no immersion medium on the cover slide, the nanoparticle is hardly seen, if the immersion medium matches the slide and the fiber probe too well, the probe vanishes optically. As the experimentalist needs to see both probe tip and particle during the picking

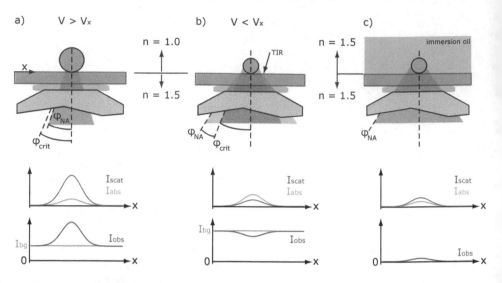

Fig. 5. Illumination schemes (top row) and scattered, absorbed, background and observable intensities for different configurations. a) A sphere of volume $V > V_x$ scatters more than it absorbs, hence $I_{obs} = I_{scat} - I_{abs} + I_{bg}$ shows a positive bump – the contrast of the particle is bright on dark. b) For particle volumes $V < V_x$ the bump is negative, as scattering dominates (dark on bright). In addition, a lens with a larger NA is applied, which increases the background I_{bg} due to total internal reflection (TIR): the red sectors of the incident light are fully reflected. c) Immersion oil on the cover slide removes the refractive index jump and hence any reflection: $I_{bg} = 0$. The ratio V/λ becomes nV/λ, so the effective wavelength in the medium is smaller. This, in turn, pushes the position of V_x towards smaller particle volumes.

procedure, a compromise is needed. Instead of a regular immersion oil, which matches common BK7 cover slides and optical fibers too well, a liquid of a lower refractive index is desired. Glycol[3] and glycerin[4] turned out to be suitable, as they are non-volatile, viscous, not too toxic, their index of refraction can be tuned down by mixing with water, and they are easily obtainable. Keep in mind that the immersion liquid between cover slide and microscope lens, i.e. *under* the slide, must be chosen according to the lens manufacturer's specifications.

2.2.3 How to select the right nanoparticle
Selecting an appropriate nanoparticle *before* it is attached to the probe tip is important, as this is the key ability of the procedure provided here. In the case of metal nanoparticles, Mie's theory helps: a single gold sphere, will back-scatter a characteristic spectrum. The peak position λ_0 is related to the diameter D, e.g., a gold sphere embedded in a liquid of a refraction index $n = 1.5$ has its peak position at

$$\lambda_0(D) = \frac{1.15 \cdot 10^9}{2.09 * 10^6 - 20D^2} . \tag{3}$$

---

[3] Ethylene glycol, [107-21-1], $n = 1.43$
[4] Glycerol, [56-81-5], $n = 1.473$

This rule of thumb, derived from (Sönnichsen, 2001), is valid for sphere diameters between 20 and 160 nm, and a powerful handle for size control (Härtling et al., 2008). A plot can be seen in Fig.4 *right*. If the particle of interest was not symmetric, its "effective diameter" would change for different orientations – the peak position is affected according to rotating either the particle or the incident polarisation (Kalkbrenner et al., 2004). As a benefit of this behaviour, the orientation of metal nanowires/ellipsoids/optical antennas can be determined (Olk et al., 2010). In addition, single nanoparticles and multiplets, despite of they can't be resolved, will produce an Airy disk of a red-shifted colour. For a well-known, reproducible species of nanoparticles, the step of spectral analysis may be omitted, as according to equation 1 an intensity analysis might be sufficient in order to identify "healthy", single nanoparticles.

For nanoparticles differing from metal scatterers, alternative identification methods must be provided, according to the actual requirements of the intended experiment.

2.2.4 Choice of glue

The adhesive needs to provide these properties:

- The nanoparticle must be fastened securely and reliably.

- The glue shouldn't alter the mechanics of the scanning probe.

- The optical properties of the glue must not interfere with the actual experiment.

For affixing metal nanoparticles to glass, APTMS[5], APTES, or PEI[6] are fine, but other material combinations may demand other specific glues - large plastic spheres can actually be pronged. Detailed procedure descriptions exist in order to obtain densely packed monolayers of these materials. In this application here, such a perfect monolayer isn't mandatory: it is sufficient to dip the probe tip into a 5-10% aqueous solution of APTMS for some tens of seconds. This allows the silane ends of the molecules to bind to the glass surface. Unbound, excess molecules can be rinsed off by a spill of purified water. As a result, a holey monolayer of amino groups is ready to bind to gold or silver.

The last requirement for the glue, being optically neutral, can turn out to be tricky. For larger particles, a patchy monolayer of molecules isn't affecting the effective medium strongly. But in a single-molecule Raman spectroscopy experiment, the Raman signal from the glue must not outshine the response from the molecule of interest.

2.2.5 Alignment and attaching

The previous sections provide a scanning probe and a candidate particle. Both tip and particle are very small, so their images are (Airy) disks, probably of different size, colour, and intensity. When setting up a regular scanning probe, the tip is positioned manually some microns over the sample surface. Now the experimentalist aligns the two image disks, frequently jumping forth and back between the two focal planes. A scaled fine drive screw for the microscope will be helpful, as the focal plane of a high-NA lens is very thin. When aligning, two (Airy) disks are centred on each other - the Rayleigh criterion plays no role here. In addition, using a "blunt" tip adds lateral tolerance, as in Fig. 1. In total, the centring tolerance is in the range of the tip radius, which in turn can be close to the (irrelevant) Rayleigh criterion.

Once tip and particle are aligned on the vertical axis, and separated vertically by just one or two micrometers, the actual picking can take place. For most setups, a regular approach procedure at moderate oscillation amplitudes is sufficient: once the tip is "parked" on the

---

[5] 3-Aminopropyl-trimethoxysilane, [13822-56-5]
[6] Polyethylenemin, [29320-38-5]

nanoparticle for a few seconds, the tip may be retracted. If everything went well, the tip *and the nanoparticle* disappeared from the focal plane - moving the microscope up will bring both back. Moving the tip laterally will quick-check if the particle is actually affixed thoroughly enough. Fig.6 is the corresponding view through the eyepiece of a microscope.

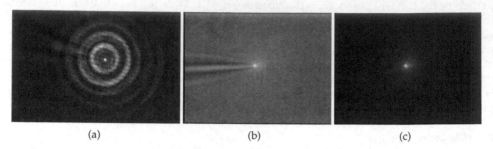

(a)   (b)   (c)

Fig. 6. Black-and-white optical images of particle-decorated fiber probes. The fiber ($n \approx 1.5$) is immersed in glycol ($n = 1.42$). *Left* The tip produces a V-shaped "shadow" to an external lamp pointing to the upper left of the image. The bright concentric rings (cones actually) stem from fiber-transmitted laser light that is coupled into the other end of the fiber. The white spot in the middle is the back-scattered light from a gold sphere of 80 nm diameter. The illumination scheme corresponds to Fig.5c). *Middle* The laser is switched off, the pointing direction of the external lamp is different now. *Right* The external lamp is switched off, the microscope's illumination is the only remaining light source. The excellent contrast demonstrates the concept of immersed fibers.

To illustrate the possible results, Fig.7 provides some electron microscopy images of particle-decorated tips. Note that an electron image is not necessary for day-to-day quality control, as the procedure, as given here, provides a rigorous check whether the particle is attached or not.

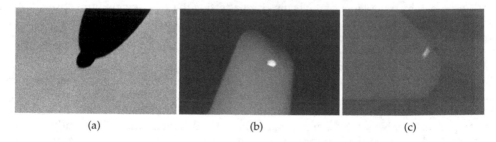

(a)   (b)   (c)

Fig. 7. SEM images of various particle-decorated tips. a) a 100 nm Au sphere, carbon coated for improved scanning electron imaging. b) Material contrast image of a 80 nm Au sphere. Despite of the broken glass tip, the decoration process worked and produced a usable particle probe. c) Material contrast image of a Ag nanorod, 125 nm in length and 55 nm in diameter. Note the odd spatial orientation of the nanorod.

If the particle refuses to follow the tip, the binding process couldn't take place for several reasons. The nature of these reasons is somewhat unclear, but pressure, time, density of the glue molecules and oscillation amplitude may play a role. Therefore, increasing the normal force by tuning the setpoint may help, as might a reduction of the oscillation amplitude. If this didn't lead to success yet, harsher methods may be appropriate: a deliberate disturbance of the distance controller may lead to the desired result. An audible clearing of the throat, coughing, clapping, and, as the last resort, tapping on the optical table (just a soft finger tip), can induce a gentle, but non-ignorable disturbance. Be aware that due to the small area of a nanoparticle, the forces on it may reach easily the magnitude of GPa, which in turn deforms the spherical particle to a heavily patted lump. While this opens the opportunity of nano-minting[7], the change of the shape leads to a visible(!) redshift of the plasma resonance.

2.2.6 Optical antennas - multiple particles

For optical antenna applications, e.g., using a metal nanostructure as a concentrator of the incident field in order to produce a high intensity, say, for Raman scattering experiments, multiple particles turn out to be advantageous, (Fleischer et al., 2008; Jiang et al., 2003; Li et al., 2003; Nie & Emory, 1997; Olk et al., 2007). For this reason, it may seem attractive to obtain particle decorated probes consisting of two (or even more) particles.

Alternatively, one may try to attach the second particle *directly* to the first one, using a specific glue for this purpose: in this case, Au is to be attached to Au, so alkane dithiols (Brust et al., 1995) or more rigid linking molecules (Pramod & Thomas, 2008) are appropriate, if a fume hood is available. A rather flexible link can be achieved by dipping the tip (which is already the first nanoparticle) for 10 minutes into undiluted decanethiol[8]. The residual thiol can be rinsed off by isopropanol. Especially if the second particle is of similar or bigger size than the first one it is recommended to passivate the residual, but unused molecule on the glass tip: for APTMS, exposure to a mild organic acid, mixed into the immersion liquid, turned out to be beneficial. In a straightforward attempt, one may repeat the procedure above and

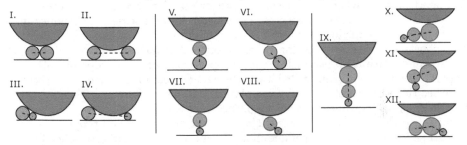

Fig. 8. Possible alignments (only two-dimensional, for the sake of simplicity) for two particles on a tip. The dotted lines connect the particle centres and depict the principal axes. *I.-IV.* If both particles are attached directly to the glass tip. *V.-VIII.* The first particle is affixed to the glass tip by a specific glue (yellow coating), while the second particle is mounted on the first one by a different, specific glue (red coating). *IX.-XII.* Illustration of some mentionable situations for three-particle chains. For small and many particles, quality control by electron microscopy is advised.

affix an additional particle to the glass tip. This will produce particle-decorations as in Fig.8 *left*. Note that in this configuration, the final orientation of the principal axis[9] is dependent on

---

[7] For comparison: a typical AFM cantilever, spring constant 40 N/m, bent by 50 nm, applies a force of 2 μN. This force, distributed over the projection of a sphere of 25 nm radius, exceeds 1 GPa, more than the yield strength of many metals.

[8] 1,10-Decanedithiol, [1191-67-9].

[9] The line connecting the two centres of the particles.

the particle size, on the distance between the particles, and even on the order of attaching the particles[10]. Generally, the principle axis will be rather horizontal than vertical.

In terms of accuracy, the concepts of Fig.3 can be transferred to this two-particle system, but note that now, the relative position, and consequently, the orientation of the principal axis, is of fundamental importance for the optical antenna effects, Fig.8 *middle*.

The situation "worsens" for particle chains of three or more particles, Fig.8 *right*. If a straight, vertical chain is desired, one has to mind the ways of how the manufacturing process could go wrong. Of course, such "faulty" assemblies may open opportunities for other usage: a folded chain might be considered as a collection of two-particle antennas, each sensitive for a different polarization orientation and/or wavelength.

2.3 Applied scanning particle probes

Two well-established application fields are Scanning Particle-enhanced *Raman Spectroscopy*, and the examination of plasmons on metal nanostructures, e.g. for fluorescence enhancement by means of *optical antennas*, and shall be discussed in detail.

Of course, the concept of a "particle on a string" is not limited to plasmonics and metal particles alone. The reader be reminded of N-V centres in nanodiamonds (Balasubramanian et al., 2008), or quantum dots Aigouy et al. (2006). Future experiments may use single nanowire lasers as a photon source (Johnson et al., 2001), or magnetic nanoparticles as a sensor (Härtling et al., 2010), not to mention antibodies or functional enzymes – the possibilities are virtually endless.

Plasmonics - optical antennas

One important field where the plasmonic properties of metal nanostructures on a tip are exploited, is the field of *plasmonics*. This denotes the discipline of engineering and manipulating metal structures and their optical properties.

An obvious field of research is the examination of the interplay among several metal nanostructures. Thanks to near-field optical coupling, the spectral properties of the metal nanoparticle on the tip is manipulated just by bringing it near to other metal structures - and vice versa. The observed effects vary from plasmon resonance de-tuning (Kalkbrenner et al., 2001; Olk et al., 2008) to four-wave-mixing (Danckwerts & Novotny, 2007).

For single particles, the enhanced near-fields are their most important property. Increasing the local intensity on a restricted volume allows for the *selective* excitation of single fluorescent molecules (Bharadwadj et al., 2007; Frey et al., 2004; Sandoghdar & Mlynek, 1999).

Raman spectroscopy

All progress in the field of enhanced near-field probes quickly found its way into the Raman spectroscopy, probably the most established application of particle-decorated tips. Due to the very small cross sections in Raman scattering experiments, the Raman community depends on high local fields, generated by Surface-Enhanced Raman Spectroscopy (SERS) and Tip-Enhanced Raman spectroscopy (TERS), using massive but pointed wire tips. The concept of optical antennas for local field enhancement was quickly embraced by the community, combined with SERS and TERS concepts, and quickly driven towards near-field enhancing metal nanostructures (Nie & Emory, 1997; Plieth et al., 1999; Stöckle et al., 2000).

The concept of using just one metal particle as an optical antenna, similar to the proposal

---

[10] Attaching a larger particle first and well on the symmetry axis of the tip will prevent a second, smaller particle from touching the glass tip.

of (Wessel, 1985), was realised early (Anderson, 2000), and has evolved towards RNA sequencing (Bailo & Deckert, 2008). Note that for scanning particle-enhanced Raman microscopy, mostly AFM-based tips with a "thin" metal film (i.e. sputtered metal coagulated to islands, which we consider as nanoparticles here) are used. Aside from the simple fact that attaching a pre-selected particle, as demonstrated here, came later into being (Olk et al., 2007), each experimentalist needs to consider the advantages and disadvantages of the two approaches: price, availability of tools, reproducibility, expected and desired local enhancement, and experimental skills play roles.

3. Demonstration - Fluorescence in the vicinity of two particles

In order to put the concepts of section 2 to a test, two different two-particle systems are analysed by means of fluorescence spectroscopy:
Fig.9 *left* illustrates the basic setup (Olk, 2008): collimated excitation light, controlled in both intensity and polarisation, is focused on the particle system, which is immersed in glycerine dyed with Nile Blue[11]. The linear polarisation of the laser source can either be rotated by a $\lambda/2$ plate or converted to so-called radially polarised light, which produces a focal polarization in the direction of z (Olk et al., 2010). Any fluorescence collected by the focusing lens is guided to a spectrometer. A typical spectrum of pure Nile Blue is in Fig.9 *right*. In a first step, the

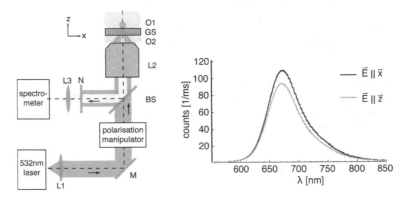

Fig. 9. *Left*. Experimental setup. L2 is a microscope lens requiring immersion oil O2 and a specific glass slip GS in order to perform optimal. O1 is the immersion medium dyed with Nile Blue. *Right*. Typical fluorescence spectra of Nile Blue for two given focal polarizations.

focal plane is moved $30\,\mu$m into the Nile Blue glycol, so the focal volume is fully in material that can fluoresce. Then, a fiber tip carrying one single Au particle of 80 nm diameter is inserted into the focus, and the fluorescence intensities are recorded for radially and linearly polarised light, Fig.10 *left*. In comparison to Fig.9 *right*, the fluorescence signal is reduced, approximately by a factor of two. This is owed to the fact that the fiber tip itself displaces the dye, so a good part of the focal volume consists of non-fluorescent glass. This is an effect that is hardly quantifiable, as the indexes of refraction are close. Usually, this is a desired advantage, but here, the similarity of the indices lets the tip vanish for the human eye. This renders it impossible to position the *bare* tip in the focus in order to quantify its displacement.

---

[11] [3625-57-8]

A different issue that can be learned from Fig.10 *left* is that for a focal z polarisation, the detected fluorescence intensity is much lower than for horizontally polarised excitation light. This can be explained by the inset in that figure: the region of high nearfields at the particle consists of two lobes capping opposite poles. For x-polarised illumination, both lobes extend into the dye, whereas for z polarisation, one lobe extends into the nonfluorescent fiber tip.

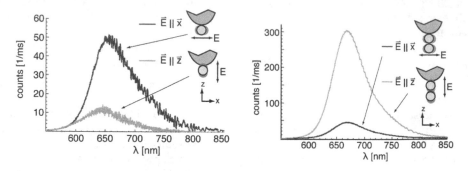

Fig. 10. Fluorescent spectra of Nile Blue for two focal polarisations. *Left* A single Au 80 nm sphere. *Right*. Two Au 80 nm spheres, the second attached to the first.

In a second step, a second Au sphere of 80 nm diameter is attached to the first one, as described in section 2.2.6. In a direct comparison, Fig.10 *right*, the fluorescence intensity is higher for both polarizations. While the signal for horizontal polarisation remained about the same, the signal for vertical polarisation increased enormously - this can be assigned to the enhanced near-field intensity between the two spheres.

A rotation check, i.e. rotating the fiber probe and the polarisation with respect to each other, provides an intensity variation of only 10%, Fig.11 hints that the two particles are close to the symmetric orientation V. in Fig.8.

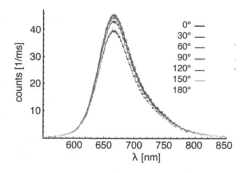

Fig. 11. Orientation-dependent fluorescence in the vicinity of two 80 nm Au spheres. The horizontal polarisation and the probe are rotated around the z-axis. The peak intensity varies by about 10%.

In a third step, the experiment is repeated with a 80 nm particle which carries a smaller particle of 30 nm diameter (both Au). A basic check with excitation light polarised along x,y,z, Fig.12 *left*, shows that the z-direction is not as pronounced anymore as in Fig.10 *right*. The rotation

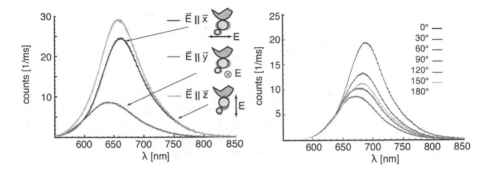

Fig. 12. Orientation-dependent fluorescence in the vicinity of a pair of 80 nm + 30 nm Au spheres. *Left.* The fluorescence intensity for x- and z-oriented excitation polarisation are rather similar, whereas in-plane y-polarisation provides a rather small signal. *Right* A rotational analysis reveals that the peak intensity varies by more than 50%.

check, Fig.12 *right*, shows a significant intensity variation for various relative orientations. This hints that the relative particle orientation is rather close to image VIII. in Fig.8, with the principal axis along x direction. In comparison to the two Au particles of equal size one realises that the principal axis here is not just slightly off the plumb line, but by a significant angle.

What's more: considering the property of the scattering cross section, equation (1), a single 30 nm particle is supposed to be nearly irrelevant in comparison to a 80 nm sphere, as the volume squares V_{30}^2 / V_{80}^2, and hence their scattering cross sections, C_{sc30}/C_{sc80}, have a ratio of less than 0.003. But thanks to near-field-mediated coupling of the two particles' plasmons, the smaller particle has a significant impact on the fluorescence in the vicinity of the particle pair. So by combining particle-decorated probe tips with a straight-forward fluorescence analysis it is possible to learn a lot about the particles and their relative orientation. This is a valuable simplification of a lab workflow: For many applications, it is already sufficient to know the direction of the projected principal axis – the tedious procedure (of mounting the tip, attaching particles, unmounting the tip keeping its orientation in mind, moving it into an electron microscope, get the particle orientation, return the tip to the optical setup while maintaining the orientation according to the SEM coordinates) is dodged. Note that for these experiments, actual *distance-controlled scanning* of the probe tip was used only for the act of mounting the particles – the orientation determination took place well off the cover slide surface.

4. Conclusion & outlook

At this point, the principles of scanning particle microscopy were explained. The technical details and "tricks" were given, and the reader should be able to follow up according to her/his own experiments. Not to use the most pointed tips available, and to exploit the properties of immersion liquids may be valuable hints.

The Application section underlined the versatility of particle-decorated probes as mobile hot spots, *i.e.* as optical antennas providing locally enhanced intensities. As an example how well the third dimension can be explored by "pulling the strings", fluorescence in the vicinity of Au nanoparticle pairs was analysed.

The given procedures are not exhaustive, but a dexterous experimentalist, after checking alternatives (Eng et al., 2010), may make use of the information here – the technology and its scientific applications are not maxed out yet, and many interesting experiments can be expected in the future.

5. Acknowledgements

The author wishes to thank Thomas Kalkbrenner for long-term inspiration on this topic, and Lukas M. Eng, Marc-Tobias Wenzel, and Thomas Härtling for providing a productive climate in the SNOM lab of the Institute for Applied Photophysics at the TU Dresden, Germany. Helge Weman of the Institute of Electronics and Telecommunications at the NTNU Trondheim, Norway, deserves gratitude for endorsing the writing of this scripture.

6. References

E. Abbe (1904). *Gesammelte Abhandlungen. Bd. 1: Theorie des Mikroskops*. G. Fischer, Jena.

M. S. Anderson (2000). Locally enhanced Raman spectroscopy with an atomic force microscope. *Appl. Phys. Lett.*, Vol. 76 no. 21, pp. 3130ff.

L. Aigouy, B. Samson, G. Julié, V. Mathet, N. Lequeux, C. N. Allen, H. Diaf, & B. Dubertret (2006). Scanning near-field optical microscope working with a CdSe/ ZnS quantum dot based optical detector. *Rev. Sci. Instrum.*, Vol. 77, 063702.

E. Bailo & V. Deckert (2008). Tip-enhanced Raman spectroscopy of single RNA strands: Towards a novel direct-sequencing Method. *Angew. Chemie Internat. Ed.*, Vol. 47, pp. 1658-1661.

G. Balasubramanian, I. Y. Chan, R. Kolesov, M. Al-Hmoud, J. Tisler, Chang Shin, Changdong Kim, A. Wojcik, P. R. Hemmer, A. Krueger, T. Hanke, A. Leitenstorfer, R. Bratschitsch, F. Jelezko, & J. Wrachtrup. Nanoscale imaging magnetometry with diamond spins under ambient conditions. *Nature*, Vol. 455, pp. 648-651.

A. Bek, R. Jansen, M. Ringler, S. Mayilo, T. A. Klar, & J. Feldmann (2008). Fluorescence enhancement in hot spots of AFM-designed gold nanoparticle sandwiches. *Nano Lett.* Vol. 8 no. 2, pp. 485-490.

P. Bharadwaj, P. Anger, & L. Novotny (2007). Nanoplasmonic enhancement of single-molecule fluorescence. *Nanotechnology*, Vol. 18, 044017.

D. Boyer, P. Tamarat, A. Maali, B. Lounis, & M. Orrit (2002). Photothermal imaging of nanometer-sized metal particles among scatterers. *Science* Vol. 297, pp. 1160-1163.

M. Brust, D. J. Schiffrin, D. Bethell, & C. J. Kiely (1995). Novel gold-dithiol nano-networks with non-metallic electronic properties. *Adv. Mater.*, Vol. 7 no. 9, pp. 795ff.

M. Danckwerts & L. Novotny. Optical frequency mixing at coupled gold nanoparticles. *Phys. Rev. Lett.*, Vol. 98, 026104.

R. C. Dunn (1999). Near-field scanning microsopy. *Chem. Rev.* Vol. 99 no. 10, pp. 2891-2927.

L. M. Eng, T. Härtling, & P. Olk (2010). Device and method for metallizing scanning probe tips. *Int. patent application*, PCT/DE2010/000579.

M. Fleischer, C. Stanciu, F. Stade, J. Stadler, K. Braun, A. Heeren, M. Häffner, D. P. Kern, & A. J. Meixner. Three-dimensional optical antennas: Nanocones in an apertureless scanning near-field microscope. *Appl. Phys. Lett.*, Vol. 93 no. 11, 111114.

H. G. Frey, S. Witt, K. Felderer, & R. Guckenberger (2004). High-resolution imaging of single fluorescent molecules with the optical near-field of a metal tip. *Phys. Rev. Lett.*, Vol. 93, 200801.

T. Härtling, Y. Alaverdyan, M. T. Wenzel, R. Kullock, M. Käll, & L. M. Eng (2008). Photochemical tuning of plasmon resonances in single gold nanoparticles. *J. Phys. Chem. C*, Vol. 112 no. 13, pp. 4920-4924.

T. Härtling, T. Uhlig, A. Seidenstücker, N. C. Bigall, P. Olk, U. Wiedwald, L. Han, A. Eychmüller, A. Plettl, P. Ziemann, & L. M. Eng (2010). Fabrication of two-dimensional Au@FePt core-shell nanoparticle arrays by photochemical metal deposition. *Appl. Phys. Lett.*, Vol. 96, 183111.

M. J. Howes & D. V. Morgan (1980). *Optical Fiber Communications - Devices, Circuits, and Systems*, John Wiley & Sons Ltd., ISBN 0-471-27611-1.

J. Jiang, K. Bosnick, M. Maillard, & L. Brus (2003) Single molecule Raman spectroscopy at the junctions of large Ag nanocrystals. *J. Phys. Chem. B*, Vol. 107 no. 37, pp. 9964-9972.

J. C. Johnson, H. Yan, R. D. Schaller, L. H. Haber, R. J. Saykally, & P. Yang (2001). Single nanowire lasers. *J. Phys. Chem. B*, Vol. 105 no. 46, pp. 11387-11390.

T. Kalkbrenner, M. Ramstein, J. Mlynek, & V. Sandoghdar (2001). A single gold particle as a probe for apertureless scanning near-field optical microscopy. *J. Microscopy* Vol. 202, pp. 72-76.

T. Kalkbrenner, U. Håkanson, & V. Sandoghdar (2004). Tomographic plasmon spectroscopy of a single gold nanoparticle. *Nano Lett.* Vol. 4 no. 12, pp. 2309-2314.

S. C. Kehr, M. Cebula, O. Mieth, T. Härtling, J. Seidel, S. Grafström, L. M. Eng, S. Winnerl, D. Stehr, & M. Helm. Anisotropy contrast in phonon-enhanced apertureless near-field microscopy using a free-electron laser. *Phys. Rev. Lett.* Vol. 100 no. 25, 256403.

K. Li, M. I. Stockman, & D. J. Bergman (2003). Self-similar chain of metal nanospheres as an efficient nanolens. *Phys. Rev. Lett.*, Vol. 91, 227402.

S. Nie & S. R. Emory (1997). Probing single molecules and single nanoparticles by surface-enhanced Raman scattering. *Science*, Vol. 275 no. 5303, pp. 1102-1106.

P. Olk, J. Renger, T. Härtling, M. T. Wenzel, & L. M. Eng (2007). Two particle enhanced nano Raman microscopy and spectroscopy. *Nano Lett.*, Vol. 7, no. 6, pp. 1736-1740.

P. Olk, J. Renger, M. T. Wenzel, & L. M. Eng (2008). Distance dependent spectral tuning of two coupled metal nanoparticles. *Nano Lett.*, Vol. 8 no. 4, pp. 1174-1178.

P. Olk (2008). *Optical Properties of Individual Nano-Sized Gold Particle Pairs*. Dissertation, TU Dresden, *http://nbn-resolving.de/urn:nbn:de:bsz:14-ds-1218612352686-00553*.

P. Olk, T. Härtling, R. Kullock, & L. M. Eng (2010). Three-dimensional, arbitrary orientation of focal polarization. *Appl. Optics*, Vol. 49 no. 23, pp. 4479-4482.

P. Pramod & K. G. Thomas (2008). Plasmon coupling in dimers of Au nanorods. *Adv. Mater.* Vol. 20 no. 22, pp. 4300-4305.

T. G. Philbin, C. Kuklewicz, S. Robertson, S. Hill, F. König, & U. Leonhardt (2008). Fiber-optical analog of the Event Horizon. *Science*, Vol. 319 no. 5868, pp. 1367-1370.

W. Plieth, H. Dietz, G. Sandmann, A. Meixner, M. Weber, P. Moyer, & J. Schmidt (1999). Nanocrystalline structures of metal deposits studied by locally resolved Raman microscopy. *Electrochimica Acta*, Vol. 44 no. 21-22, pp. 3659-3666.

W. Rechberger, A. Hohenau, A. Leitner, J. R. Krenn, B. Lamprecht, & F. R. Aussenegg (2003). Optical properties of two interacting gold nanoparticles. *Opt. Comm.* Vol. 220 no. 1-3, pp. 137-141.

V. Sandoghdar & J. Mlynek (1999). Prospects of apertureless SNOM with active probes. *J. Opt. A: Pure and Appl. Opt.* Vol. 1 no. 4, pp. 523-530.

C. Sönnichsen (2001). *Plasmons in Metal Nanostructures*. Cuviller, Göttingen.

N. Stephenson (1996). Mother Earth Mother Board. *Wired*, Vol. 12, ISSN 1059-1028.

R. Stöckle, Y. D. Suh, V. Deckert, & R. Zenobi (2000). Nanoscale chemical analysis by tip-enhanced Raman scattering. *Chem. Phys. Lett.* Vol. 318, pp. 131 - 136.

E. Udd (1990). *Fiber Optic Sensors*, Wiley Interscience, ISBN 0-471-83007-0.

B. Uhlig, J.-H. Zollondz, M. Haberjahn, H. Bloeß, & P. Kücher (2007). Nano-Raman: Monitoring nanoscale stress. *AIP Conf. Proc.*, Vol. 931 pp. 84-88.

M. T. Wenzel, T. Härtling, P. Olk, S. C. Kehr, S. Grafström, S. Winnerl, M. Helm, & L. M. Eng (2008). Gold nanoparticle tips for optical field confinement in infrared scattering near-field optical microscopy. *Opt. Express*, Vol. 16, pp. 12302-12312.

J. Wessel (1985). Surface-enhanced optical microscopy. *J. Opt. Soc. Am. B*, Vol .2 no. 9, pp. 1538-1541.

F. Zolla, G. Renversez, A. Nicolet, B. Kuhlmey, S. Guenneau, & D. Felbacq (2005). *Foundations Of Photonic Crystal Fibers*, Imperial College Press, ISBN 1-86094-507-4.

Permissions

The contributors of this book come from diverse backgrounds, making this book a truly international effort. This book will bring forth new frontiers with its revolutionizing research information and detailed analysis of the nascent developments around the world.

We would like to thank Dr. Moh. Yasin, Professor Sulaiman W. Harun and Dr Hamzah Arof, for lending their expertise to make the book truly unique. They have played a crucial role in the development of this book. Without their invaluable contribution this book wouldn't have been possible. They have made vital efforts to compile up to date information on the varied aspects of this subject to make this book a valuable addition to the collection of many professionals and students.

This book was conceptualized with the vision of imparting up-to-date information and advanced data in this field. To ensure the same, a matchless editorial board was set up. Every individual on the board went through rigorous rounds of assessment to prove their worth. After which they invested a large part of their time researching and compiling the most relevant data for our readers. Conferences and sessions were held from time to time between the editorial board and the contributing authors to present the data in the most comprehensible form. The editorial team has worked tirelessly to provide valuable and valid information to help people across the globe.

Every chapter published in this book has been scrutinized by our experts. Their significance has been extensively debated. The topics covered herein carry significant findings which will fuel the growth of the discipline. They may even be implemented as practical applications or may be referred to as a beginning point for another development. Chapters in this book were first published by InTech; hereby published with permission under the Creative Commons Attribution License or equivalent.

The editorial board has been involved in producing this book since its inception. They have spent rigorous hours researching and exploring the diverse topics which have resulted in the successful publishing of this book. They have passed on their knowledge of decades through this book. To expedite this challenging task, the publisher supported the team at every step. A small team of assistant editors was also appointed to further simplify the editing procedure and attain best results for the readers.

Our editorial team has been hand-picked from every corner of the world. Their multi-ethnicity adds dynamic inputs to the discussions which result in innovative outcomes. These outcomes are then further discussed with the researchers and contributors who give their valuable feedback and opinion regarding the same. The feedback is then collaborated with the researches and they are edited in a comprehensive manner to aid the understanding of the subject.

Apart from the editorial board, the designing team has also invested a significant amount of their time in understanding the subject and creating the most relevant covers. They scrutinized every image to scout for the most suitable representation of the subject and create an appropriate cover for the book.

The publishing team has been involved in this book since its early stages. They were actively engaged in every process, be it collecting the data, connecting with the contributors or procuring relevant information. The team has been an ardent support to the editorial, designing and production team. Their endless efforts to recruit the best for this project, has resulted in the accomplishment of this book. They are a veteran in the field of academics and their pool of knowledge is as vast as their experience in printing. Their expertise and guidance has proved useful at every step. Their uncompromising quality standards have made this book an exceptional effort. Their encouragement from time to time has been an inspiration for everyone.

The publisher and the editorial board hope that this book will prove to be a valuable piece of knowledge for researchers, students, practitioners and scholars across the globe.

List of Contributors

K. S. Lim and H. Ahmad
Photonic Research Center, University of Malaya, Kuala Lumpur, Malaysia

H. Arof
Department of Electrical Engineering, Faculty of Engineering, University of Malaya, Kuala Lumpur, Malaysia

S. W. Harun
Photonic Research Center, University of Malaya, Kuala Lumpur, Malaysia
Department of Electrical Engineering, Faculty of Engineering, University of Malaya, Kuala Lumpur, Malaysia

Atsuhiro Fujimori
Saitama University, Japan

Vladimir Demidov, Konstantin Dukel'skii and Victor Shevandin
S.I. Vavilov Federal Optical Institute, St. Petersburg, Russia

Paloma R. Horche
ETSI Telecomunicación, Universidad Politécnica de Madrid, Spain

Carmina del Río Campos
Escuela Politécnica Superior, Universidad San Pablo CEU, Spain

Shin-ichi Todoroki
National Institute for Materials Science, Japan

Xavier Artru and Cédric Ray
Université de Lyon, Université Lyon 1, CNRS-IN2P3, Institut de Physique Nucléaire de Lyon, France

Paulo Antunes, Fátima Domingues, Marco Granada and Paulo André
Instituto de Telecomunicações and Departamento de Física, Universidade de Aveiro, Portugal

Toto Saktioto
Advanced Photonics and Science Institute, Faculty of Science, Universiti Teknologi Malaysia, Skudai, Johor, Malaysia
University of Riau, Pekanbaru, Riau, Indonesia

Jalil Ali
Advanced Photonics and Science Institute, Faculty of Science, Universiti Teknologi Malaysia, Skudai, Johor, Malaysia

Sinead O'Keeffe, Denis McCarthy and Elfed Lewis

Optical Fibre Sensors Research Centre, University of Limerick, Ireland

Dan Sporea and Adelina Sporea
National Institute for Laser, Plasma and Radiation Physics, Laser Metrology Laboratory, Romania

Phillip Olk
NTNU Trondheim, Norway

Printed in the USA
CPSIA information can be obtained
at www.ICGtesting.com
JSHW011437221024
72173JS00004B/845